HEAD OF BUDDHA
(Gandhara—*Museum of Fine Arts, Boston*)

Gautama Buddha

IQBAL SINGH

Delhi
Oxford University Press
Bombay Calcutta Madras

Oxford University Press, Walton Street, Oxford OX2 6DP
Oxford New York
Athens Auckland Bangkok Bombay
Calcutta Cape Town Dar es Salaam Delhi
Florence Hong Kong Istanbul Karachi
Kuala Lumpur Madras Madrid Melbourne
Mexico City Nairobi Paris Singapore
Taipei Tokyo Toronto
and associates in
Berlin Ibadan

Originally published by Boriswood, London 1937
Published in India with corrections and
additional matter 1994
Second impression 1994

ISBN 0 19 563455 1

Printed at Raj Bandhu Industrial Co., New Delhi 110064
and published by Neil O'Brien, Oxford University Press
YMCA Library Building, Jai Singh Road, New Delhi 110001

APROPOS THE SECOND IMPRESSION

WRITTEN more than five decades ago, this book was first published less than a year before Munich. Later much of the stock was apparently destroyed in the Blitz. Since then it has been out of print and even second-hand copies have been hard to come by. That, in a sense, may be claimed to establish the need for a reprint.

Needless to add, it was the product of a certain mood and youthful enthusiasm for the subject which went with that mood. The mood could scarcely be expected to endure after all those years; nor, for that matter, the sustaining enthusiasm. While preparing the text for a new impression, inevitably, I have been increasingly aware of many a fault-line in the treatment of the theme. There has been a strong temptation at times to rewrite, or at least revise, it radically in the light of all the second, and perhaps wiser, thoughts that have suggested themselves. That would certainly have the advantage of making the argument considerably less vulnerable.

On serious reflection, however, the temptation had to be resisted—and not necessarily because of the great expense of time and effort which such a revision would entail. It had to be resisted, primarily, because to make any structural changes in the basic argument dictated by hindsight would be a kind of vandalism. The virtues of disenchanted hindsight which comes with age are not to be underrated. But it may be doubted if, in the ultimate analysis, they are an adequate substitute for the prodigal, if fleeting, insights accessible to the mind while it still has the courage of its innocence and insouciance, and has not yet gone on to the defensive. Whatever its vulnerability, therefore, it seems best to leave the original structure unchanged, except for some relatively minor corrections and the incorporation of some additional material in the last section of the book dealing with the vicissitudes suffered by the Buddha in a changing world.

There are at least two other and related questions which may legitimately be asked. The first is whether the interpretation offered here of Gautama's world-view and the processes through which it was reached, retains sufficient residual validity to warrant republication of the book after such a long interval. The second and more pertinent question is whether the personality of Gautama and his understanding of the nature of human experience and destiny connect significantly with the deeper preoccupations and problems of our tormented age at any point. As regards the second question, there can be little doubt in any dispassionate mind that the answer has to be in the affirmative. After the vast and insensate hecatombs witnessed during the Second World War, the gas-chambers of Auschwitz and Buchenwald, the wanton atomization of Hiroshima and Nagasaki, and the prolonged agony of Vietnam, to say nothing of the continuing ministry of nuclear fear and terror under which we all live like so many somnambulists, it cannot reasonably be argued that the word of the Buddha has lost its relevance. Quite the contrary would seem to be true, whether or not anybody cares to heed his word.

The first question, obviously, is not for the present writer to answer. But some clarification may not be out of place. This tentative essay in biographical interpretation was never intended for the edification of the faithful. And for the good reason that its essential concern is not with faith or cults which have fastened themselves upon the Buddha; its concern is with the man in all his complex but ennobling humanity. If some fervent devotees of the Buddha found my approach unacceptable, to put it mildly, this was hardly surprising.

What came as a surprise—and it was an agreeable one—was that among the younger generation of Buddhists, especially in Ceylon (or Sri Lanka as we know it today), the book evoked not only some interest, but a sympathetic response. There was no feeling, at any rate, that I had presented the Buddha, or his view of the human condition and how it may best be comprehended, to disadvantage.

It only remains to be added that I had not conceived my task as a trespass into the wide but rather jealously guarded field of Buddhist scholarship either to impress or challenge the specialists—*genus irritabile* as Professor A. Foucher has somewhat uncharitably categorized them in his *Life of the Buddha According to the Ancient Texts and Monuments of India*. For this reason it was not considered necessary to load the book with footnotes, recondite references and cross-references, or diacritical marks which pass for status symbols among the scholarly establishment. Nor am I now persuaded that anything would be gained by making good this deficit in the second impression. A famous mid-nineteenth century French translator of Indian classics, Hippolyte Fauche, in the introduction to his translation of the *Gita Govinda* explained sensibly: *Mais nous avons jugé superflu que la traduction marchât avec un grand bagage de notes; elles sont inutiles pour quelques uns; peut-être seront-elles pour d'autres un peu insuffisantes*. . . . I feel much the same way. An academic disguise is unlikely to make this work more acceptable to the experts; and it may well put off the common reader, if such there be. So it must travel light and unencumbered by any scholarly ballast as, indeed, did the Tathagata himself during his long ministry of compassion.

Delhi, Autumn 1993 I.S.

PREFACE

FOR the past two thousand years or more the figure of Gautama, the Buddha, has attracted hagiographers and legend-makers. It need not imply disparagement of their efforts if one admits that for the most part they leave one with a sense of frustration. This frustration is the more keen because the Buddhist canon unmistakably appears to point to an experience of a very crucial nature, and one cannot help feeling that a proper understanding of that experience would lead to a clearer realization of some of the more urgent human problems. In this resides the justification for another book about the Buddha.

I should like to add that I do not presume to have discovered any final or definitive clue to Gautama's personality. Perhaps the time for such discovery has not yet come. In the present work, therefore, I have done no more than restate some of the issues which needed restatement; shift the emphasis from origins to attitudes—and from attitudes to those subtle processes of feeling and apprehension of which these attitudes were by-products; and finally, offer an interpretation of the Buddhist legend compatible with common sense.

There is one point where I have departed somewhat radically from the accepted canons of hagiography. I have not treated Gautama as an isolated and prodigious phenomenon. For so to regard a man is to do him less than justice. I have, therefore, dealt at some length with the background of the age in which Gautama lived. It was in every way an important and eventful epoch; it saw not only the rapid rise and fall of great empires, but the crystallization of new patterns of human outlook; and much that then happened is still relevant to us.

It is possible that some people may find this book lacking in those particular graces which come only from an attitude of pious reverence towards the subject. I am not unaware of that deficiency, and would crave indulgence. Ultimately, however, a critical approach is perhaps the highest compliment one can pay—even to a Tathagata. And I need hardly avow my intellectual sympathy for Gautama's view of life: to lack sympathy with that view is to lack sympathy with what seems to be most sane and normal in humanity.

I.S.

CONTENTS

APROPOS THE SECOND IMPRESSION page 3

PREFACE 6

PART I

THE WORLD OF THE BUDDHA 11

PART II

BIOGRAPHICAL 43

I Maya's Dream and the Nativity of the Buddha 45

II The Republic of Kapilavastu 54

III Some Early Events 77

IV The Pursuit of Pleasure 95

V The Crisis and Renunciation 118

VI After Strange Gods 142

VII Enlightenment 160

VIII Turning the Wheel of Doctrine 177

IX Growth of the Order 187

X Years of Wandering and a Day 208

XI Fools in the Order 226

XII The Great Decease 233

XIII Posthumous Postscript 249

PART III

THE WORD OF THE BUDDHA 251

I The Wheel 253

II The Way 262

III The Void 271

PART IV

AND THE WORD WAS MADE FLESH 289

PART V
THE BUDDHA IN A CHANGING WORLD 305

EPILOGUE 355

LITTLE TO ADD 359

ACKNOWLEDGEMENTS 382

BIBLIOGRAPHY 384

INDEX 392

ILLUSTRATIONS

HEAD OF BUDDHA *frontispiece*
(Gandhara–*Museum of Fine Arts, Boston*)

BODHISATTVA OF THE BLUE LOTUS *facing p.* 95
(Ajanta–*The Director of Archaeology, Hyderabad*)

WOOD NYMPH 128
(Sanchi–*Archives of the Musée Guimet*)

BUDDHA IN SHRINE 163
(Ajanta–*The Director of Archaeology, Hyderabad*)

DEATH OF BUDDHA 233
(Swat–*Copyright, Archaeological Survey of India*)

LANDSCAPE 286
(attributed to Ma Kuei. From *Civilizations of the East* by René Grousset–*Hamish Hamilton*)

LISTENING TO MUSIC 293
(attributed to Chou Fang. From *Introduction to Chinese Paintings* by Arthur Waley–*Ernest Benn)*

TO
EDITH

PART I

THE WORLD OF THE BUDDHA

The watchman said, The Morning cometh, and also the night: if ye will enquire, enquire ye: return, come.

Isaiah

THE transition from the preclassical to the classical, and in a larger sense, from the ancient to the modern world, is marked by the slow, faltering, and yet unmistakable emergence of a new human attitude—the attitude vaguely described as humanism. The significance of this subtle change can best be realized by contrasting it with the attitude which preceded it, and of which, however incompletely, it represents a negation.

It would of course be historically untrue to suggest that the preclassical period of human history stands out as a single and uniform epoch. Within its boundaries—which extend on the one hand to the remotest sources of recorded time, and on the other encroach ominously upon our own—we can discern several distinct gradations of mental and social make-up, corresponding to different stages of human evolution, each having its own peculiarities. But distinct as no doubt these phases are from one another in matters of detail, nevertheless they seem to show a fundamental psychological identity, a common moral basis of integration. And though in the actual process of concrete manifestation this can be seen to throw out widely varied patterns, determined by the conditions of time and place, throughout each phase there runs the unbroken chord of MYTHISM.

In the last resort, whatever their individual characteristics, the whole series of brilliant cultures which flourished in the pre-classical period derived their strength from the Myth-Complex. The history of these cultures, therefore, can only in an indirect sense be regarded as human history. The real force that sustained them, their common denominator as it were, was the Myth in one or other of its infinite representations; and it is hardly surprising that they should have risen and fallen with the particular myth on which they happened to be based.

The Myth-Complex naturally excludes the direct exercise of human [intelligence. For man must either believe in himself or in the Myth. He must act according either to the light of his own reason, or to the dictates of an imaginary agency outside himself. The ancient world gave a preponderating influence to the Myth in the conduct of earthly affairs; and in doing so it virtually denied all initiative to humanistic thought, subordinating reason to mystery. Strictly speaking, the mythogenetic tendency in man is the legacy of his pastoral and primitive-tribal past. To primitive man the Myth was an absolute necessity. Its underlying assumption, the striking incomprehensibility of phenomena, was quite natural, even inevitable, to his state. In fact, it was something more than an assumption to him: it was the most outstanding fact of his experience. Finding himself utterly at the mercy of mysterious elements, he felt the need for setting up a mediating and protective agency. In other words, he created the Myth, to lighten 'the burden and the mystery' of this unintelligible world, which seems to have weighed on him as heavily as the shadow of death itself, and to give a semblance of order and meaning to the bafflingly hostile universe. But it has been a common occurrence in the world of men for intermediaries to usurp the place of the principals; image-makers have often ended by becoming slaves to the images they themselves have made. Thus not only did the Myth-Complex survive long after the conditions which had given rise to it had ceased to exist, but with time it actually became even more deep-rooted in human imagination. Such is the force of habit.

As the early centuries wearily succeeded one another, man made slow but definite advance towards civilized life. Gradually he ceased to be a nomad wandering across arid wastes, and settled down to the infinitely more fruitful occupation of farming, and by so doing added such highly ingenious arts and crafts to his accomplishments as irrigation and agriculture, cattle-breeding and pottery-making, weaving and house-building, working gold and copper, the making of clothing, the brewing of beer, the use of arithmetic, and the devising of calendars. Already in the aeneolithic age, in the fertile valleys of the Nile, the Indus, and the Euphrates, we find him properly organized for urban life, possessing the written word in the shape of a pictographic script, constructing houses with tiled bathrooms, and devoting minute attention to the principles of efficient plumbing and drainage. Tormented by recurrent hunger, he was driven across vast and difficult distances, from the grey northern skies to the sun-baked plain of the Ganges. Goaded by desire for gold, his hazardous self-imposed quest led this insatiable prospector from Northern Africa to the furthermost extremity of Asia, from the shores of the Dead Sea to Land's End. By degrees, the horizon of his social consciousness widened. He was persuaded to transfer his allegiance from the tribal to the kingly ideal, to march to death behind a flag instead of a totem-pole. In consequence, powerful dynasties came into being, great monarchs were born. In Egypt, by some happy accident, he discovered that the green copper ore malachite, which had long been considered an indispensable boudoir-requisite by Cleopatra's countrywomen, and was employed by them as a cosmetic, could, through the alchemy of fire, be transmuted into a beautiful metal much more useful than gold. The discovery was epoch-making. With it began the age of metals. Large-scale industrial production became possible, and pyramids sprang up to break the monotony of desert stretches and preserve the mummified corpses of the mighty.

The earth continued to perform its unceasing gyrations.

Ur of the Chaldees rose to fame and prosperity, developed a most elaborate mercantile code, inaugurated a system of cheques, bills of lading and exchange, invoices, and many other technical conventions of commerce. The Phoenician merchant-adventurers sailed forth to the ends of the earth with their rich cargoes of exquisitely coloured textiles which Hecuba was proud to wear next to her delicate skin; glassware as finely tinted as the Bohemian and Venetian glass that we know; and mass-manufactured metallurgical wonders to suit all pockets and all tastes. They established 'factories' at various important ports along the Mediterranean, and probably as far West as the Channel Islands, the 'Tin-Islands' of the bronze age. And wherever they went, besides their manufacture, they carried with them the alphabetic mode of writing.

On the strength of the humble shell-fish, the *Murex trunculus* and the *Purpura lapellus*, Tyre and Sidon quickly grew into the world's most important cities of export. As precious as the excreta of the silkworm, the tiny colourful drop of fluid pressed out of the purple mussel was turned into dyes of many shades, ranging from soft crimson to the richest ruby, from a delicate amethyst to the dark red of coagulated blood. The sons of Canaan made huge fortunes from their jealously guarded trade-secret, and having exhausted home fisheries sought out fresh sources of this profitable raw material on alien shores, for a long time enjoying a virtual monopoly of the sea-borne trade between Asia Minor and Europe. The fame of their nautical knowledge and artistic achievements spread all over the Asiatic world and the Near West; kings and princes vied with one another to requisition the services of their professional seamen, master-craftsmen, and technicians. Hiram, 'a son of a woman of the daughters of Dan and a man of Tyre,' a creative genius as great and versatile as Leonardo da Vinci and Michael Angelo, provided the temple of Solomon with its ornaments and decorations—the two pillars, Jachin and Boaz, the tri-coloured veil, the molten sea supported on twelve oxen, the wreaths, the pommels of the chapiters, the lavers, the basons, the candlesticks, the censers, and indeed most of the costly sacerdotal property of the house of the Lord.

Stars continued to trace their fixed courses through empty space; and but for an imperceptible retrogression of the equinoxes and some trivial geological disturbances, the precarious cosmic equilibrium was somehow maintained. Under the exhilarating influence of *Soma,* the Indian mystagogues in their peaceful retreats among verdant Himalayan valleys meditated on the great question-marks of birth and death. Less mystically inclined, an amorous son of David found all the delights for which he craved in dwelling on his beloved's breasts which were as 'clusters of the vine'; and wrote love's sweetest song. The disillusioned and disconsolate Preacher, also claiming to be a son of David, on the other hand, ruefully proclaimed that love, laughter, joy, wealth, wine, wisdom, knowledge, matrimony, parenthood, kingship, and all the innumerable states of man under the sun were at the root only weariness and vexation of spirit, profiting nothing; that all things passed, yet everything repeated itself like the wind that 'whirleth about continually.'

However, the bronze age, once over, was not to return to this planet. Once man had learned how to temper iron, he put it to a far more serious purpose than making such household necessities as sewing- and knitting-needles for his women to display their skill. The new weapons of destruction designed by him greatly quickened the tempo of world events. The glorious days of military heroes began. Vast armies chased each other across the earth's surface at the command of ambitious war-lords. Feudal upstarts fought among themselves for land and power. More rapidly than ever before, kingdoms were lost and won by the sword. Yet, through all these dramatic transformations, the Myth, the omnipotent, omnipresent spirit of the Myth, remained supreme. Man, the quintessence of dust, the prodigal child of nature, accepted his thraldom humbly, unquestioningly. We find him prostrate before stone altars; offering his own blood, and the blood of those whom he hated or loved, to satisfy the whims of unseen, vindictive gods; defending himself desperately against non-existent demons. We find him abjectly worshipping the sun and moon, the creatures of the air, the monsters of the undersea-gloom, crawling insects and the fleet-footed horse, the hay-eating calf and beasts of prey, the winged dragon of heaven and 'the pretty worm of Nilus that kills and pains not,' the lizard and frog, the cat and the mouse, the ghosts of his distant dead, certain prominent parts of his own body, and diabolical death-masks and nail-fetishes of his own making. But with the advent of classical times we become conscious of a change in human outlook. The still small voice of protest against the ridiculous spectacle of man's idolatry begins to be audible. Faint and tremulous, it rises through the broken cries and mournful lays of lachrymose priests and prophets, above 'the smoke of praise going from ocean rim to rim,' and over the countless sacrificing hands 'upraising the chalice flowing to the brim' to quench the thirst of strange gods and even stranger goddesses. Faint and tremulous, but with a new sense of conviction—even of hope.

The tendency in man to think rationally, to reduce the chaotic universe of his sense-impressions and intuitions to a coherent and logical order, is much older than the Hellenes.

Hellenism itself was not altogether a new birth: partly, it was the rebirth of the old Minoan-Mycenaean civilization whose spark, though it had suffered prolonged eclipse during the Dark Ages between the thirteenth and the eighth centuries B.C., had never completely been extinguished, and was to rekindle with added brilliance on the achievement of the Ionian revolution. As Elliot Smith says, most classical writers have underestimated the extent of the Greeks' indebtedness to those who preceded them. Erdmann's assertion that philosophy began with the Greeks contains about as much truth as the biblical legend, according to which, before the Almighty bestirred himself—in the year 4004 B.C.—to accomplish the herculean task of separating the earth from the waters, there was nothing but darkness upon the face of the deep. Our sources of knowledge bearing on man's earliest attempts to find an adequate answer to the riddle of creation and the mystery of his own being are no doubt extremely limited and fragmentary. But there is enough historical evidence to discredit the ignorant and preposterous view held by Erdmann that among the Eastern peoples, with the solitary exception of the Jews, thought does not attain to a sufficiently high degree of self-consciousness to comprehend its own nature and intrinsic dignity. Even the naïvely lyrical hymns of the Rig-Veda, congeries of half-formed myths or crude allegories as they are, at times attain a note of profound and sober reflection pointing to a highly developed faculty for abstract thought. Indeed, one of the hymns belonging to the later period of the Vedic Age offers an explanation of the First Cause, which comes nearer to the Aristotelian conception of the 'Unmoved Mover' than to the anthropomorphic bed-time story given in Genesis. It reads:

'There was then neither what is nor what is not, there was no sky, nor the heaven which is beyond. What covered?

Where was it, and in whose shelter? Was the water the deep abyss [in which it lay]?

'There was no death, hence was there nothing immortal. There was no light [distinction] between night and day. That One breathed by Itself without breath, other than It there has been nothing.

'Darkness there was, in the beginning all this was a sea without light; the germ that lay covered by the husk, that One was born by the power of heat [tapas].

'Love overcame It in the beginning, which was the seed springing from mind, poets having searched in their hearts found by wisdom the bonds of what is in what is not.

'Their ray which was stretched across, was it below or was it above? There were seed-bearers, there were powers, self-power below and will above.

'Who then knows, who has declared it here, from whence was born this creation? The gods came later than this creation, who then knows whence it arose?

'He from whom this creation arose, whether he made it or did not make it, the Highest Seer in the highest heaven, he forsooth knows, does even he not know?'

It is little exaggeration to say that this hymn contains a startling presentiment of almost every basic problem round which metaphysical enquiry has revolved for the past two thousand years and more. Of course, it will serve no useful purpose to import into this passage any ultra-modern interpretations. As it stands, it can in fact be interpreted to imply a number of completely contradictory principles. The obvious conclusion to be drawn from its ambiguity is that the poet is himself lost in the confusion of his own thought, is striving in vain to arrive at luminous certitude. That he should not be able to give us an unequivocal answer as to the ultimate reality is inevitable. But, as Max Müller who translated it comments, the remarkable thing about this statement is that it signifies a complete break with the prevailing notions of physical deities with clearly pronounced sexual and other personal attributes. Its historic interest consists in the fact that for such a bold departure from 'a personal to an impersonal or rather superpersonal cause' to have become possible there must have been a long and previous philosophic incubation. Thus in the very midst of a flourishing animistic religion, we encounter individual spirits courageous enough to have ventured as far as the border line between mythology and metaphysics.

This precocious metaphysical activity was not confined to India. The Hindus were by no means the only people among the ancients to have spent restless nights speculating on the nature and origin of the world. The cosmic questions postulated by the early Indian thinkers, which, at a subsequent date, were to be reiterated with fresh emphasis by the 'seven wise men of Greece,' had troubled inhabitants of many other lands as well. In the richly fertile delta of the Nile, the votaries at the shrines of Osiris had, from the very dawn of history, been preoccupied with finding a satisfactory solution for the problem of cosmogony. Similarly, the esoteric broken and single lines of the *Yih-King*—a compilation which goes back to the closing centuries of the second millennium B.C.—seem striving to hint at what was to form a subject of fashionable controversy among the wise and handsome Athenians, and was to give rise to a most noble and accomplished literature. The book itself was intended mainly for the purpose of furnishing oracular clues somewhat on the lines of Hesiod's *Works and Days*, *Old Moore's Almanack*, and the prophetic columns of our Sunday newspapers; princes and peasants alike had recourse to it in order to seek positions of the constellations, favourable or unfavourable to a proposed undertaking. Surprisingly, however, we find in it the uneasy germ of an idea which is the very essence of dialectics: the idea of the 'permeation of opposites.' *Yi*, represented by the sign of the sun placed in opposition to that of the moon, was meant to symbolize the eternal principle of change and fluctuation through an unbroken continuity. The book enumerates a whole series of 'polar concepts' such as Yang and Ying, heaven and earth, light and dark, male and female, strong and weak, etc.; and the implication is that the constant interaction of these opposite categories determines all developments both in the macrocosm and microcosm. The underlying theme of the *Yih-King* can, therefore, be understood as anticipating, though still within the framework of a metaphorical convention, not only Heraclitus' conception of flux and Zeno's paradoxes, but also to some extent the everlasting strife of the Hegelian thesis and antithesis, which every instant brings forth new syntheses.

It is true that all these philosophic intuitions of antiquity are still inextricably involved in the tentacles of myth and metaphor. But everything under the sun has to be void and formless before it can be moulded into shape. The bond of what is, must be sought in what is not. From the womb of nebulous eternities are born stars, sharp like unto crystals. And the human mind, too, had to pass through a transitional phase of diffuse, incoherent mysticism before it could achieve the precision of a dissecting-knife. In these early, half-hearted, hesitant experiments at rational interpretation of the content of experience, we can observe the way being paved step by step for that great intellectual revolution which began with the classical age, and which is even now very far from being complete. Out of this inchoate mass of thought-substance was to arise a new critical approach, not merely to the world without but also to the world within.

It is a difficult and futile undertaking to set out to fix chronology where psychological changes of this kind are concerned. In any case, events are more important than dates; and it is best to point to certain landmarks. One such landmark is furnished by the breakdown of the Assyrian hegemony in the near East. In or about 606 B.C., Asshur fell. Of the exact manner of this fall many accounts are given, based partly on legend and partly on fact. But one thing is beyond doubt: the wound sustained by Asshur was grievous and fatal. His fall, says Ragozin, had been brought about by his own weight, his own wickedness and folly. After a long and terrible siege lasting nearly two years, Nineveh, the pride of Assyria, was laid waste under the combined thrust of the Median and Babylonian forces, till there was hardly any one left to lament her ruin. Betrayed by his own trusted nobles, deceived by his own subjects, in despair, Saracos set fire to his palace and together with many of his concubines perished in the flames. Though Asshur had wings he was not of the phoenix kind: he never rose again from his ashes. The prophecy contained in 'the burden of Nineveh' was thus literally fulfilled. Only, it had come to pass far too late to be a source of satisfaction and joy to the prophet. Long before Nineveh was razed to the ground, Nahum, the Elkoshite, had died. Moreover, whatever else came out of it, the collapse of Assyrian power certainly brought no relief to the long-suffering body of Judah. As has happened practically throughout history, for Judah—alas!—it was once again merely a choice between two evils of equal magnitude. The iron heel of Babylon, which came down upon her almost as soon as the ruthless Assyrian whip was lifted, had no more mercy in it. Nebuchadnezzar, the king of Babylon, had many fine virtues, but love for his neighbours was not among them. He did not take any more sympathetic view of Judah's legitimate aspirations for national independence than the Assyrian tyrants had done before him; with a firm and powerful hand he crushed the ill-advised rebellion led by Zedekiah, king of Jerusalem, who paid the price of his foolhardy adventure first with his eyes and later with his life. Indeed, the merchants and patriarchs of the Holy City would have fared far better had they heeded Jeremiah's unheroic but extremely sensible advice, instead of having the unhappy prophet lowered into a foul-smelling dungeon, and threatening his life daily because he, a patriot, persisted in telling the bitter truth, warning the Jews that, in the circumstances, discretion was by far the better part of valour.

But what lends significance to the dissolution of the Assyrian Empire—the last of the pre-classical empires—is not the negligible bearing it had on the tragic destiny of the Chosen People, but a number of vastly more important historical considerations. There followed in its wake a most crucial redistribution of the balance of power in the Levant and the Middle East, which in those days, like Western Europe to-day, constituted the vital nervous centre of the civilized world. Its repercussions were felt even in such comparatively remote lands as China and India. In fact, the realignment of political forces which ensued upon the disintegration of Assyria was no arbitrary or accidental phenomenon. If we analyse its causes, we find a continuous process leading up to this climax, and it is impossible to avoid the conclusion that the actual shifting of power was but the outward sign of a highly revolutionary, sociological transition.

It has already been remarked that by the end of the seventh century, the change-over from bronze to iron had been effected in most of the countries which then formed the cradles of civilization. This technical revolution was to transform the face of earth. On the one hand, it opened fresh possibilities for the satisfaction of man's acquisitive itch; on the other, it raised commodity-production to a hitherto undreamed-of level. These two factors could not but have tremendous repercussions in social, moral, and intellectual spheres. There were, in fact, serious crises in the social and political structures. The existing social systems may, or may not, have been fully aware of the precise nature of their predicament; but they were faced with nothing less than the necessity for fundamental readjustments. In some ways the dilemmas confronting the ruling classes of that age were akin to those which confront the capitalist oligarchy to-day. Unwittingly, almost against its will, the old tribal-cum-patriarchal feudalism was compelled to seek for itself a new basis. The acute need for revising the existing social codes seems to have been felt everywhere. In India, China, and Greece law-givers were called in to devise all kinds of new sanctions to meet the situation. The ultimate equilibrium was established on a basis very different from the original pattern. Solon's social reforms give some idea of the transformations in economic relationships characteristic of that age. In the political sphere two antithetical expedients were resorted to: feudal imperialism and feudal democracy. The prolonged conflict between these two systems is familiar history; it ended in victory for imperialism. The world had not yet become safe for republicanism, even of a very restricted and aristocratic kind.

The break-up of Assyria naturally gave Babylon the supreme political power in the Near East. But the Babylonian supremacy was fated to be short-lived. Nebuchadnezzar was as imperious and accomplished an oriental potentate as any, and though prone to excessive dreaming, had great military gifts. He conquered many lands and crowned his triumphal march by inflicting a crushing defeat on Necho in Egypt. He was, we are told, a most chivalrous and loving husband. To compensate his Median wife, Amytis, for the loss of her native hilly scenery, tradition credits him with having set to work his vast army of slaves on the construction of an artificial hill in the very heart of Babylon. The Greeks counted this hill among their seven wonders of the world, and called it 'paradeisos'; for it was here that Nebuchadnezzar laid the fabulous hanging gardens with their beautiful terraces, the multi-coloured flower-beds, the herbaceous borders, the stately avenues of exotic trees, the shady bowers, and the summer-house where the capricious queen might always enjoy peace and pure air. Apart from this, his interest in town-planning is known to have been something more than a mere dilettante pursuit. Babylon the Great owed much of its architectural beauty to his inspiration; and Daniel records the pride he took in his handiwork when he represents him as exclaiming: 'Is not this the Great Babylon, which I have built for the Royal dwelling-place, with the might of my power and the glory of my majesty?' But neither his power nor the glory of his majesty were to outlive him. With the murder of Belshazzar the Babylonian sceptre passed into Median hands, which in their turn, were to relinquish it to the Persians. And the rising star of Cyrus was to throw all other luminaries of the contemporary political firmament into shadow.

The rise of Persia is the central fact of the period with which we are concerned—the age of Buddha. In comparison all other events fade into insignificance. Its tremendous historical influence has frequently been stressed. 'The Persian,' writes G. Buchanan Gray, 'is vastly more than a mere successor to the Median Empire. With the Medes the Aryans first took a conspicuous place in world-history; but it is their kinsmen, the Persians, who first became a world-power.' What gives the rise of Persia to supremacy a singular interest, is that it was not a slow growth: it was a sudden and spectacular event. The highly dramatic process of expansion was completed 'within the space of a single generation . . . by a series of conquests which followed one another with a rapidity scarcely equalled except by Alexander, and by the Arabs in the first generation after the death of Mohammed.' The whirlwind military campaigns of the Persian forces drove the wedge of Achaemenian authority right to the shores of the Mediterranean in the West, bringing under its yoke the Asiatic Greeks, the Chaldeans, the Egyptians, the Phoenicians, and many other peoples of the mainland; in the East they planted the Persian flag on the banks of the Indus, thus extending the periphery of Persian hegemony to the North-Western provinces of India. In less than thirty years the Persian territories exceeded 'that ever obtained by the greatest monarchs of Mesopotamia and the Nile Valley.'

The Asiatics are generally slow to be moved into action, but once they get under way they become the very fiends of energy. At the time when Gautama, already a middle-aged man, was trying with little measure of success to show his countrymen the way of compassion and loving-kindness, and Confucius was engaged on the thankless task of re-establishing the crumbling patriarchal feudalism in China on a Utopian basis through the 'rectification of names,' another oriental worthy, Darius, was busy putting the final touches to his preparations for the first historic attack of Asia on Europe. His well-trained, mechanized expeditionary columns were soon to traverse Asia Minor, cross the Bosphorus, and even the Danube, by means of an improvised bridge of boats in 516 B.C.

For the next few generations the course of both Eastern

and Western politics was determined by the Persian initiative. In the West the Persian invasion was instrumental in drawing Athens, until then a second-rate city-state, out of its obscurity into a position of unrivalled prominence, thus enabling it to play its unique rôle in the history of civilization. The Confederacy of Delos, a league composed of a number of small independent states pledged to mutual assistance, came into being primarily for the purpose of keeping the Persian fleet at bay, to secure the independence of the member-states, and to safeguard their common maritime interests in the Aegean, which formed the basis of prosperity of the Greeks of the archipelago as well as of the peninsula. In this 'Little Entente' of the Aegean, Athens, by virtue of its strategic situation and long sea-going traditions, naturally came to exercise a dominating influence. When the Persian advance was checked at Marathon by the brilliant strategy of Miltiades and Polemarch Callimachus, Athens secured a victory the moral value of which transcended the mere military gain. For Marathon was mainly an Athenian triumph, won by Athenian arms, and without much outside help. As such, it not only filled the citizens of Athens with 'national pride and sanguinary patriotism,' and gave them 'high ideas as to their power and destiny,' but enhanced the prestige of Athens in the eyes of the whole Greek world. The heroes of Marathon turned the Athenian supremacy in the Aegean League into an undeniable fact. Paradoxically, Marathon was not only the making of Athens: it also led to the ultimate disruption of the Athenian Republic. It intensified still further the acute rivalry between Athens and Sparta, and sowed the seeds of a tension whose poisonous harvest was reaped in the shape of Peloponnesian wars, which proved to be the death-bed throes of Greek republicanism and the civilization it had created.

In the East the Persian Empire had subtler, more indirect, but none the less decisive influence. The Egyptian Pharoahs, we know, had claimed divinity for their own persons. Whether for reasons of modesty or good sense, the Achaemenian sovereigns were content to perfect the doctrine of kingship by divine grace. Indeed, they did something more; they made it work in practice as it had never worked before. The unification of extensive dominions stretching from the Indus to the Nile—an area as large as, and in some ways richer than, Russia—which these energetic earthly representatives of Ahuramazda successfully carried out, created an attractive precedent for feudal imperialism; and it is hardly surprising that this ideal should have become increasingly popular during the next few centuries.

The time was ripe for imperial adventure. Tribal republics and feudal principalities, emasculated by prolonged internecine struggles, were tottering all round. Both in China and India, warlike princes found splendid opportunity to play at being emperors; and after a number of abortive attempts two of them succeeded in their aim. Immediately after the death of Alexander, Chandra Gupta brought the best part of India under his yoke, and founded the house of Mauryas. In China the Empire came a hundred years later with the foundation of the Han dynasty. The new type of feudal organization replaced the old social order in the Eastern as well as the Western world. Empires multiplied; conqueror after conqueror followed in the footsteps of Cyrus the Great, who may be considered the prototype of all future military leaders.

Widespread as the ramifications of the Persian Empire were, from a political point of view, the part it played in the cultural sphere was even more important. Not that the Persians themselves contributed anything outstanding to the cultural heritage of mankind. The Achaemenians were not, in fact, an intellectually brilliant people. Their actual achievement in the realms of art, literature, and thought amounts to little; and what they achieved is, even at its best, mediocre and derivative. Empires have performed, however, a vital historical function, the value of which cannot be over-emphasized: they have been the vehicle of cultural diffusion without which a great deal of ideological evolution would never have been possible. This rôle the Persian Empire fulfilled admirably—probably even better than some of the empires which came after. The Persians were temperamentally not particularly creative, but possessed a saving grace: they were without any deep-seated prejudices or traditions, and therefore easily assimilated the different cultures with which they came in contact in the course of their victorious march through many lands. Not being born islanders with acutely developed insular traits and habits of mental and social inbreeding, they felt no need for carrying their little bit of Persia with them when they went to the outposts of their Empire. Instead of shutting themselves up in air-tight enclosures in remote garrison-towns in order to keep their tribal inhibitions warm, the Persian administrators freely mixed and intermarried with the populations over which they ruled, absorbed their culture and gave them their own culture in exchange. Indeed, they were so highly impressionable that Herodotus contemptuously remarks that he never knew a people more amenable to adopting the customs and manners of the countries where they happened to go. But the characteristic which Herodotus deprecated was precisely what gave the Persian Empire its great historical significance, and made the Persians extremely useful as cultural links between peoples who had no direct means of communication.

Moreover, their very mediocrity served them in good

stead when it came to the question of setting up an efficient administrative machinery for the Empire. 'Not only in extent did the Persian Empire far surpass any that had gone before,' says the *Cambridge Ancient History*, 'but in the organization to which it was subjected; it may indeed be regarded as the first attempt to bring a large number of different races and nationalities into a single government which assured to the whole the rights and privileges as well as the burdens and responsibilities of members of the state.' But more crucial by far than the expedient and profitable conception of an Imperial Commonwealth, willing regularly to pay tribute to the Emperor in solid gold, which the Persian rulers evolved, was the recognition on their part that the maintenance of such a far-flung Empire depended entirely on the existence of adequate lines of communication, allowing the utmost mobility to royal troops. The classical historians are unanimous that both Cyrus and Darius were aware of the urgency of the problem of inter-imperial transport. The arterial routes which they etched over their territories, we are told, 'achieved the completeness and excellence of the communications of the later Empire of Rome.' Ctesias in the missing section of his Persian History is known to have given a detailed account of the trunk-route linking Bactria and India. Herodotus mentions several other routes, and in particular describes at length the one connecting Susa with Sardes in Lydia, which had excellent inns and posting-stations provided at convenient stages along the route, where 'messengers mounted on swift horses stood always in readiness to carry the king's letters and despatches so that these passed as fast as the horse could travel without delay by day or night,' accomplishing a journey which normally took three months in less than a week.

If the rise of Persia as the premier world-power is the

most outstanding political event of the sixth and fifth centuries B.C., then the elaborate network of roads between various parts of the Empire which the Persian monarchs created, is the greatest achievement of their rule. The sociological importance of the vast improvement in communications which they effected, no doubt to serve their own imperial ends, cannot be exaggerated. It marked an advance comparable to the revolution in transport that followed the introduction of the compass into navigation much later in human history; and we may reasonably assume that it had similar social and cultural consequences. The new road system did not merely afford quick passages for the army and royal mail: it opened many fresh channels for the flow of trade, made freer migrations and movements of population possible, and carried wealth and the light of civilization to a large number of communities who had hitherto dwelt in primeval darkness, being completely cut off from the outside world. The process of civilization was further greatly accelerated by various important public utilities which the Achaemenid kings inaugurated in order to better the economic conditions of their subjects, increase the sources of taxation, and stimulate large-scale exchange of commodities. In this category we must mention the completion by Darius of the canal linking the Nile, a little above the city of Bubastis, with the Red Sea at the point where modern Suez stands: a project which, nearly a century earlier, Necho had contemplated, even started, but abandoned. Until the later Egyptian kings let this waterway fall into disuse and decay, it provided water-communication between India, Persia, Egypt, and the Mediterranean, which, it is not difficult to imagine, had a very considerable commercial and strategical value.

Susa, the original capital of the Great Kings of Persia, never became a great and beautiful city in the sense in which Memphis and Cnossus, Ur and Sidon, Nineveh and Babylon were great and beautiful cities. In architectural design, in regal splendour, in formal loveliness, and even in matters of drainage it could not rival the capitals of some of the earlier Empires. But it surpassed them in its truly cosmopolitan character, the profuse intermixture of different racial elements represented in its floating population, and the picturesque human spectacle it presented. At once a point of confluence for a great many cultural streams and a centre for the dispersal of civilization, with roads converging upon it from all directions, it easily became the metropolis of the East.

From all parts of the world, interminable trains of caravans arrived here to get rid of their heavy loads or on their way to one of the twenty satrapies into which the Empire was divided for administrative purposes. Hoping to make quick fortunes, merchants from far-off shores journeyed to this international mart, bringing with them raw materials and manufactured articles, luxuries and foodstuffs of all descriptions. In the bazaars of Susa the produce of practically every land from India and China in the East to Egypt and Hellas in the West could be procured; sparkling fabrics and intricately worked jewellery, stones which burned like stars, pearls which might have been eyes, glittering beads and shells that echoed the undying murmur of the sea-drift, trinkets and toys, cosmetics and toilet requisites, rich spices, ingeniously cast girdles of chastity, electrifying aphrodisiacs which could multiply the pleasures of love-making for youthful hedonists and even rekindle the flame of passion in the veins of the aged and infirm, intoxicating beverages distilled from the stems of mountain plants and medicinal gums extracted from Arabian trees. Male and female slaves were auctioned in the crowded live-stock markets alongside the camels and dogs: the buyers were offered a wide choice in tint and texture of flesh, ranging from the curly-haired Ethiopian boys with dark bodies to the virgins captured from the northern wilds, their bosoms white as snow, their hair like beaten gold. The enigmatic Chinaman rubbed shoulders with the mysterious Magi; the contemplative Indian listened to the noisy children of Israel; the athletic, handsome Greek exchanged greetings with the top-heavy and highly inscrutable Mongol.

A multitude of tongues combined to form the babel that was Susa. Visitors to this city learned many useful things; most important of all, they learned that the world was bigger than they had conceived, that indeed it was a world without end. They communicated to each other their rudimentary conceptions of terrestrial geography; some maintaining that the earth was like an exact circle drawn by a pair of compasses with ocean belts engirdling it, others ridiculing this fantastic notion and arguing that our planet was foursquare, with Asia and Europe just the same size. They discussed contemporary politics and spoke of revolts and uprisings in Ionia. In particular they must have dwelt on such matters as vitally concerned their interests; for instance, the heavy import and octroi duties levied by the State. There were also, we may suppose, murmurs of discontent against the rigorous censorship exercised by the king's official over private correspondence passing along the roads, all of which was subjected to such minute scrutiny that when Histiaeus, who was staying at the Persian Court, wanted to send an important confidential message to Aristagoras in Miletus, he could find but one safe way: 'which was by taking the trustiest of his slaves, shaving all the hair off his head, and then pricking letters upon the skin, and waiting till the hair grew again.'

It was an eventful age: happenings followed one another

in rapid sequence. From the middle of the sixth century B.C. until the rise of Alexander, Susa was the clearing-house for all the news of the world. And there was a great deal of news: momentous developments succeeded one another: Athens had risen, and Babylon had fallen; Egypt was conquered and Jerusalem sacked. Necho's Phoenician flotilla, which had sailed into the Southern Ocean by way of the Erythrean Sea, had skirted the Libyan coast 'until the sun was on their right hand,' and after three hard winters had doubled the Pillars of Hercules. The exploits of Skylax, the Carrian sailor, who, under orders of Darius, sailed down the Indus, navigated a part of the Great Ocean 'where all the waters meet,' found his way into the Red Sea, and arrived safely in the vicinity of Suez, had something of the publicity accorded to the voyages of Diaz, Columbus, and Vasco da Gama, at the end of the fifteenth century of the Christian era, and such topical interest as a non-stop, round-the-world solo flight would have to-day. Legends were legion of mariners who had set out to sea in search of fame and fortune, and been lured to watery graves by heartless sirens inhabiting lonely and uncharted isles. The professional scribes mixed history and fiction till it was impossible to distinguish one from the other. Travellers and explorers brought with them stories of hair-raising adventure; they gave vividly coloured accounts of all the strange things they had seen, and perhaps even more, of things which they had invented. Reports were current of a happy race of Hyperboreans dwelling in a paradise of peace and plenty, and ferocious cannibal tribes infesting the wild hills on the Indo-Persian border; of fabulous 'gold-digging ants larger than foxes' and birds with rainbow-tinted plumage found in the luxuriant tropical forests.

Doubtless, among the visitors to this cosmopolitan centre, there were men possessed of more serious intent, their minds preoccupied with deep and earnest religious or philosophic quests. In the caravanserais and hostelries provided for travellers, under the brooding stillness of brittle Eastern nights, when each little breath or rustle of wind takes on an enormous and uncanny significance, these strangers sometimes whispered to one another their peculiar notions of the Good, the True, and the Beautiful. They affirmed their faith in the one great God, a jealous God, in the habit of appearing to the faithful in terrifying visions of a whirlwind, a cloud of brightness, 'a fire infolding itself,' a magnificent sapphire stone; or they asserted the existence of a vast Pantheon of gods and spirits, each representing some element of the earth, and the sky, and the water. They spoke of fierce and angry prophets who, on the slightest provocation, threatened everybody around with instantaneous annihilation in the most gruesome and violent fashion, and of prophets so gentle and compassionate that they had nothing but words of praise even for their mortal enemies; of the priests who accepted only the best meat and wine as sacrificial offering, and of those who managed to subsist on a frugal diet of carrots and water, and enjoined similar abstinence upon all those who came to them for ministration; of wizards who read the future into people's dreams, and alchemists who were reputed to know the art of turning base metals into noble ones. They talked of Wise Men who denied God altogether, and others who were so amorous of him that they looked upon their life on earth as a painful exile, and wandered about the world completely desolate because 'the alien garment' of flesh made them feel vagabonds from their heavenly beloved; of diverse forms of worship and adoration-mysteries practised by men; of cruel rituals enacted in honour of the pitiless deity of destruction, and the voluptuous ceremonies to celebrate the festival of the lovely goddess who had risen from soft sea foam imbued with all the melting tenderness of a love-sick woman. Half-understanding, they discussed such abstruse things as time and eternity; and some of them maintained that nothing was stationary under the sun, while others repudiated the very possibility of motion.

Some among them were almost certainly aware of the different theories of life and after-life prevailing among those who had been initiated into these profound mysteries; they could probably explain how according to the teachings of certain schools of adepts there was not one but many tedious cycles of births and deaths to be suffered before one could hope to attain the final extinction; and how others held that death was a dreamless sleep from which there was no awakening, that with the return of dust to dust there was an end to the pain and sorrow of humanity, and that this being so, blessed were the dead and dying. Thus, ideas and beliefs originating in the minds of eccentric and enlightened individuals who lived outside the furthest limits of the Empire managed slowly to percolate into the imperial city, and were in turn broadcast to the outlying parts of the Empire.

There were events of another kind. While Ezekiel was wrathfully inveighing against the whoredoms of Aholah and Aholibah, Thales had quietly been measuring the height of the Pyramids by the shadows they cast. Pythagoras, besides founding a society of whimsical vegetarians, discovered the properties of numbers. The edifice of an Euclidean universe was rising in the human mind curve by curve, line by line. In China, the laughing philosopher Lao Tzu managed to write *Tao Te King* during his sporadic public appearances, and preached the most extraordinary wisdom; Confucius contented himself with the less ambitious task of teaching human beings better manners, and died of a broken heart; and Tang Hsi 'conjured with words till wrong seemed right and right wrong.' In India, Kapila, and Gautama after him, caused considerable confusion among the Brahmans.

Man's universe was thus expanding, not merely in a geographical and economic sense, but also was becoming intellectually and psychologically more comprehensive. In the wake of purely material developments, there followed a corresponding widening of his mental orbit. The rapidly shifting political and social relationships had their spiritual counterpart. Even looking at it in historical perspective and comparing it with other periods of revolutionary changes of human outlook, the intellectual and emotional ferment which began towards the end of the seventh century B.C. appears to have been of a magnitude which has hardly ever since been surpassed, and is equalled only by the ideological reorientations that came as an aftermath to the discovery of the New World and the sea-route to the East by a number of West European adventurers during the fifteenth and the sixteenth centuries of our era. Indeed, the avenues of thought and feeling which the men of Western Europe began dimly to realize only after the Middle Ages had been explored with a surprising thoroughness by some of the Eastern and the Near Western nations two thousand years earlier; and the giant figures which so impressively stride across the pages of Renaissance and post-Renaissance history assume humbler, more natural proportions when considered beside the representative personalities of the classical epoch.

During the two centuries which elapsed between the fall of Asshur and the collapse of the City State in Greece there was many-sided progress in man's physical environment. It naturally brought with it a new sense of confidence and certitude as well as fresh energy and enthusiasm. A keen spirit of scepticism and enquiry was abroad; and everywhere attempts were made to penetrate the mist of magic and superstition and to arrive at a more objective explanation of reality. What had so far been accepted as revealed truth on the mere word of an interested priestly class, or out of those atavistic fears which are embedded in the depths of human memory, was now examined under the ruthless light of reason. The experiment was carried out on a most comprehensive scale. Practically every existing belief was questioned and every aspect of experience subjected to minute and rational analysis. Iron had given man the necessary strength and courage to challenge the supremacy of the Myth; for the first time he openly declared himself as the measure of his world. Thus the germ of humanism was conceived.

The whole process, of course, was full of extremely complex developments. It had manifold ramifications; and to represent it simply as an effervescence of the human will caused by newly discovered power, would be to miss its infinitely subtle nuances. It actually worked at times in a direction antithetical to what might logically have been expected. If, on the one hand, some men were carried away by a great wave of elation and joy at the prospect of having broken free from bondage to the mythic monitor, on the other the vistas unfolded by the new analytical vision left them somewhat unhappy, bewildered, and very much perturbed by the condition of humanity. The void created by the breakdown of the Myth produced a sense of futility and frustration, a feeling of despair at having taken on a burden too heavy to bear. The acute degree of self-consciousness achieved by the age imposed upon the more sensitive type of individual an overwhelming psychological strain; and some of the most alert minds of the time felt overwhelmed. It is a general characteristic of all significant transitions of the human spirit that they develop on contrapuntal lines; and side by side with the growth of self-confidence in the classical world, we hear a wistful, elegiac note of despondency. Out of this mood of mingled exultation and anguish, of audacity and hesitation, arose the tragic view of life.

It was a highly infectious view and came to be shared by people as far removed from one another in space as the tragedians of Periclean Greece and the anonymous early Buddhist poets. In its attempt to understand its own nature, obscure and unsuspected abysses were revealed to the human soul; and it is hardly surprising that under the stress of these discoveries, to live or to die should have become an agonizingly immediate question.

This was not an entirely new dilemma. Long before Job, in his profound misery, cursed the day he was born, and the Preacher exposed the worthlessness of all those things on which human beings build their happiness and hope, long before Gautama made the fact of suffering the pivot of his scheme of things, and Empedocles taught the students in his academy to look upon man's body as his tomb, in ancient Egypt 'a man weary of life' had spoken of death in endearing terms, and pined for the 'happy fate of those who are over yonder.' 'Death is before me to-day,' he had sung, 'like the convalescence of a sick man, like going forth after an illness. Death is before me to-day like the smell of myrrh, like sitting beneath the sail of the boat on a breezy day. Death is before me to-day like the longing of a man to see his home, when he has spent many years in captivity.' This melancholy and somehow mysteriously satisfying sentiment achieved its most perfect and refined expression in classical poetry and thought.

The psychological crisis of which we have spoken was not confined to any one country; it was of universal character. The vast convulsion which shook the mind of man from its protozoic lethargy did not stop at any racial or geographical boundaries. It is not merely a coincidence that, almost simultaneously, Gautama in India and Thales in Miletus rejected the hieratic presumptions and laid the foundations of a reasonable view of life. We do not know for certain whether they knew of each other; and it would be a rash undertaking to attempt to determine the extent to which the world-view of one shows the influence of the other. Further, we have no means at our disposal to settle the priority dispute between them. But such questions have little meaning for any one whose concern is with human values. For us, therefore, the important thing—and one which is absolutely beyond doubt—is that they partook of the same spiritual travail of humanity.

The movement of intellectual awakening which began with them remains one of the most earnest efforts ever made by man to realize his own manhood. Some time towards the end of the eighth century B.C. a Hebrew prophet, still very much muddled, but possessed of greater enlightenment, and more passionately wise than all his tribe, had posed a peculiar question. 'Watchman,' he had asked vehemently, 'what of the night?' The reply which came was equally strange and enigmatic. 'The morning cometh,' the Watchman said, 'and also the night: if ye will enquire, enquire ye: return, come.' Within the next few generations, however, the point of this baffling statement was to become somewhat more clear. For though the night was to persist, as it in fact persists to-day, there were men born upon this earth who, to all appearances, might have been heralds of dawn, pilgrims of light. One such man was Gautama.

PART II

BIOGRAPHICAL

. . . Nothing extenuate,
Nor set down aught in malice.

Othello

I

MAYA'S DREAM & THE NATIVITY OF THE BUDDHA

There was a Birth, certainly . . .
T. S. Eliot: *The Journey of the Magi.*

AT the time of Gautama's arrival on earth the State had not yet taken upon itself the function of keeping a careful record of births, marriages, and deaths. And if Lytton Strachey was right in maintaining that 'ignorance is the first requisite of the historian—ignorance, which simplifies and clarifies, which selects and limits, with a placid perfection unattainable by the highest art,' then Gautama is an ideal subject for the historian. It cannot be said of him that 'we know too much about him' to write about him. For, in fact, we know next to nothing about his person with any degree of certainty. Our knowledge of him is largely conjectural, being almost entirely derived from a mass of legendary and devotional literature in which he figures as the hero.

However, uncertainty of this kind is a double-edged weapon. If it is conducive to 'a placid perfection unattainable by the highest art' on the one hand, on the other it leads to controversy and doubt. In the case of Gautama, for instance, the lack of conclusive information about his life has raised considerable speculation as to his historical authenticity. Antiquarians like M. Senart have been inclined to deny that there is any historical basis to the Buddhist legend. Instead of taking the view that 'the Buddha' is a myth which has grown round the personality of Gautama, they prefer to argue the reverse, holding that 'Gautama' represents an attempt to actualize the myth of the Buddha, who is none other than the sun-hero 'issuing from the dark womb of Maya like the sun from the clouds of the night.'

Such arguments are logical enough as far as they go. The only positive answer to these sceptical scholars would be to produce Gautama's birth-certificate; and it is unlikely that any amount of historical research would make that possible. The Republic of Kapilavastu kept no birth-registers. This being so, we can only point out the danger of adopting a purely logical position in matters like this. If Gautama's birth-certificate is wanting, so too are the birth-certificates of Socrates, Jesus, and many other significant personalities of antiquity. We are not absolutely certain even about the historicity of Shakespeare—a figure much nearer to us in time than Buddha, Socrates, and Jesus.

There are considerable divergencies between the traditions current in various Buddhist countries with regard to the exact date of Gautama's descent from heaven. Rhys Davids and others have placed it in the year 563 B.C. This date may be accepted as fairly accurate (it is calculated on the basis of the tradition that Asoka's consecration took place 218 years after Gautama's attainment of Nirvana). Once again, the fact of his birth appears to be much more important than the date of his birth.

The exuberant fancy of the legend-makers has adorned the simple and natural event of Gautama's conception and birth with a wealth of miraculous and intimate detail which is sometimes embarrassing. Not only have they written up a most sensational account of 'the Wondrous and Marvellous Events' which led up to, and culminated in Gautama's birth, but they have been at great pains to fabricate an entertaining saga outlining his pre-natal adventures. However, it is no part of our undertaking to trace Gautama's career through his innumerable previous incarnations: this highly eventful story is available in the Jataka Commentary and the *Lalita-vistara* where it is presented in a manner that can hardly be improved upon. One life is enough. And this one began, the Buddhist scriptures record, in the womb of Maya, one of the two wives of Suddhodana of Kapilavastu.

Legend-makers have always and everywhere shown an incorrigible weakness for virgins: heroes and saviours are always being born of them. It is not surprising, therefore, that as with Jesus, so with Gautama, a virgin birth should have been claimed. The story that 'a virgin gave birth to Buddha from her side' seems to have enjoyed widespread currency in the Near and the Middle East during the early centuries of the Christian era. It is mentioned by St Jerome, who probably came to know of it from a study of the *Acts of Archelaus*—a work which records a controversy supposed to have taken place between the arch-heretic, Mani, and a certain Bishop Archelaus—which speaks of a predecessor of Mani, by name Terebinthus, 'who called himself Buddha' and claimed that he was born of a virgin.

However, the attempt to represent Gautama's mother as a virgin is Western rather than Indian, both in origin and inspiration. St Jerome, and Terebinthus before him, were actually labouring under a misunderstanding. The original legend does not lay any stress on Maya's virginity, but merely suggests an asexual, immaculate conception within the confines of a perfectly happy wedlock. The *Story of the Lineage* gives the following account of how Gautama was conceived:

'At that time in the city of Kapilavastu the festival of the full moon day of the month Asalha (June-July) had been proclaimed, and many people celebrated it. Queen Maya from the seventh day before full moon celebrated the festival without intoxicants and with abundance of garlands and perfumes. Rising early on the seventh day she bathed in scented water, and bestowed a great gift of 400,000 pieces as alms. Fully adorned, she ate of choice food, took upon herself uposatha vows [vows of continence], entered her bedchamber, lay down on the bed, and falling asleep dreamt this dream: the four great kings, it seemed, raised her together with the bed, and taking her to Himalayas set her on the Manosila tableland of sixty leagues beneath a great sal-tree seven leagues high, and stood on one side. Then their queens came and took her to the Anotatta lake, bathed her to remove human stain, robed her in heavenly clothing, anointed her with perfumes, and bedecked her with divine flowers. Not far away is a silver mountain, and thereon a golden mansion. There they prepared a divine bed with its head to the east, and laid her upon it. Now the Bodhisattva became a white elephant. Not far from there is a golden mountain, and going there he descended from it, alighted on the silver mountain, approaching it from the direction of the north. In his trunk, which was like a silver rope, he held a white lotus, then trumpeting he entered the golden mansion, made a rightwise circle three times round his mother's bed, smote her side, and appeared to enter her womb. Thus when the moon was in the lunar mansion Uttarasalha, he received a new existence. The next day the queen awoke and told her dream to the king . . .'

With slight variations this story occurs in other accounts of the event. It is obviously not a realistic description of what actually transpired at the time of Gautama's conception. It has, nevertheless, a great deal of psychological interest. It reveals that it is not so easy to 'remove the human stain': after describing a beautiful trajectory in the rarified atmosphere of poetic fancy, the hagiographers land by a strange irony in those very physiological regions which they wished so much to avoid. For even those who are not strict followers of the Freudian doctrine can clearly see that the legend is permeated with the suggestion of the copulatory act. The voluptuous imagery and metaphors strike an unmistakable sexual note. Even the excited flow and sequence of words tend to rise to an ejaculatory rhythm. The tree seven leagues high, the fabulous tableland, the silver mountain, the golden mansion, and the white lotus are all recognizable symbols of the organs involved in the mystery of conception and generation. Finally, there can be no misgivings as to the identity of 'the white elephant,' which, with an orgiastic flare of trumpets pierces Maya's right rib and enters her womb, even though through a reversal not at all unusual in the world of dream and myth, the whip-like tail has become a trunk 'like a silver rope.'

It is not altogether inconceivable that originally Gautama's mother herself was responsible for giving rise to the legend. It is a fairly common thing for women under the influence of eroto-mythic ecstasy to dream of phenomenal conceptions. Mothers of 'Great Men' have frequently claimed divine intervention in the conception of their children. Alexander's mother Olympias, for instance, is known to have driven Philip to distraction—and eventually into divorcing her—by persisting in her claim that she slept with snakes, and that it was Zeus-Ammon, not Philip, who begot Alexander on her, by descending into her body on her wedding night in the shape of a thunderbolt.

However, whether the legend of the white elephant was an invention of Gautama's own mother or his latter-day canonical biographers, it contains a certain amount of historical truth. The representation of Suddhodana as the king and Maya as his queen is, of course, purely conventional. But it is probably true that Maya celebrated the full-moon-day festival with a liberal distribution of alms. As elsewhere, so in India, festivals of the moon have always

d

been closely connected with fertility rituals. There would appear to be every reason for Suddhodana's wife to observe the festivals of the moon with a more than common piety. Suddhodana was anxious to have a son and heir. For this purpose he had married twice into the same family, but apparently without tangible results. He was advancing in years and the barrenness of his wives was causing him anxiety. Like every true Hindu, Suddhodana no doubt believed that his salvation depended on his capacity to produce a son. As such, he would have insisted on Maya's performing all the purification ceremonies enjoined by the Blessed Law-Giver for such occasions. What followed can easily be imagined. The identities of the 'divine bed,' of which the legend speaks, and of the conjugal bed would seem to converge into one another: the act of conception itself could hardly have been quite so immaculate as is suggested in the *Story of the Lineage*.

Under normal circumstances, for reasons of propriety, it is generally considered fit to observe a discreet silence about the crucial period between conception and birth. Not so in the legend. The price of greatness is loss of privacy: the legend does not lose sight of the Bodhisattva even while he was residing in that most private of all places—his mother's womb. During the nine months of Maya's pregnancy, we are told, the Bodhisattva was open to view 'with all his limbs and complete sense organs . . . like a beryl jewel, pure, noble, eight-sided, excellently worked, and threaded with a blue, yellow, red, white, or yellowish thread.'

Apparently truth pursues the legend-makers with a vengeance. They may make frantic efforts to fly from reality. They may take unlimited poetic licence and even ask us to believe that all the time Maya was bearing the Bodhisattva 'like oil in a bowl,' she felt no sickness whatever and was all along 'happy with unwearied body.' And yet they cannot escape physiology. It is there, inexorably—an obsessive memory. Even the colour-scheme they adopt in their description is fascinatingly physiological, giving a strangely faithful picture of the twilight that envelops the embryo in the womb.

If the legend were to be believed, it seems there was a veritable riot of prodigies at the time of Gautama's birth. Among other things, he is reputed to have emerged from the sheath 'clean, unstained with liquid, unstained with phlegm, unstained with blood, unstained with any filth.' We cannot help being surprised when a little further on the legend adds that all the same it was necessary 'for two streams of water to fall from the sky, one of cold and one of hot, wherewith to perform the washing for the Bodhisattva and his mother.' However, we need not dwell on this contradiction, nor on the other signs and portents which appeared both at the moment of Gautama's nativity and immediately after it. What is important to notice is the painful, nostalgic obsession of Buddhist poets with the physical details. At the root of this obsession there is, of course, the negative but none the less powerful yearning of the disembodied spirit for the body—even for the impurity and ugliness which goes with the body. Whereas Hebrew literature rings with a passionate craving for things of flesh and blood, Buddhist literature is permeated with an equally passionate, equally binding aversion to them. And yet the ultimate result is the same in either case: both the Buddhist poets and the Hebrew prophets seem to derive a tortured delight from gloating on matters carnal. But of this peculiarly infectious nostalgia, we will have more experience when we come to treat of the Buddhist art.

The actual facts bearing on Gautama's nativity can be

stated very briefly. When the time of his birth came, Maya, following an ancient and highly practical custom (which persists in India to this day), decided to go to her parents' home in Devadaha, a minor township not far from Kapilavastu. Every arrangement was made to ensure a safe journey for her. However, she had evidently miscalculated her time, or perhaps the strain of the journey proved too much for her: while on her way, in a grove called Lumbini grove, where the party had broken journey to rest, the throes of birth came upon her with unexpected importunity. The Bodhisattva, it seems, was in such a precipitous haste to be born that he did not give his mother even time to lie down: she was delivered of the child standing. The mother and child were brought back to Kapilavastu amidst the rejoicings of the family-circle and townspeople. But Maya herself was unable to participate in these rejoicings; she did not recover from her travail, and died a week after Gautama's birth.

The legend-makers take a light and rather complacent view of Maya's death, as though it were the most natural thing to have happened. They even suggest that, having given birth to the Bodhisattva, there was nothing left for his mother to do but die. Such, they insist, are the requirements of the Good Law. However, as many other mothers, who have had nothing whatever to do with bringing forth would-be Buddhas, are known to die under similar circumstances, the canonical explanation does not sound convincing. The cause of Maya's death was probably some complication of the after-birth. It must be remembered that she had her first and last pregnancy exceptionally late in life; she was, indeed, well over forty when Gautama was born; and the pangs of child-birth experienced by her would be much more severe than in the case of a woman in her early youth. Moreover, the lack of proper attendance must have rendered her ordeal still more painful. Mystical reasons alone would not have sufficed to cause her death.

Suddhodana had lost a wife, but that was no great matter. He had got what he most desired—a legitimate son and heir. His well-being in the hereafter was assured; he could now look forward to eternity with equanimity; and there is little doubt that, in spite of his bereavement, he felt pleased with himself. Perhaps, if he had had an inkling of his son's future, he might have been less inclined to congratulate himself.

Of course, the Buddhist birth legend does furnish a remote parallel to the story of Simeon. It speaks of the visit of sage Asita to Suddhodana immediately after Gautama's birth. There exist a great many variants of this incident. The version of the *Lalita-vistara* shows us Asita beholding the infant Buddha, detecting in him the thirty-two marks of greatness. These marks include flat feet, jaws like a lion's, blue eyes, seven convex surfaces, and a sheathed member. Seeing these signs Asita predicts: 'If he dwells in a house, he will become a king, a universal monarch . . . But if he goes forth from a house to a houseless life, he will become a Tathagata, loudly proclaimed, a fully enlightened Buddha.' He then sighs deeply and gives way to tears, because he could not hope to survive and see Gautama turn the supreme Wheel of the Doctrine, 'good in the beginning, good in the middle, good in the end, complete in the letter and the spirit, whole, pure . . .'

The story of Asita appears to be an attempt at being wise after the event, though there is nothing improbable about Suddhodana having had the horoscope of his son drawn by some astrologer of that name.

II

THE REPUBLIC OF KAPILAVASTU

SUDDHODANA enjoyed a very considerable social standing and prestige in Kapilavastu, but he did not possess the status of a monarch, as has frequently been represented. The convention of depicting Gautama as the son of a powerful potentate, as also the *Mahasudassana Sutta* in which he himself appears under the honorific title of 'the Great-King of Glory,' are of a later origin. In the first instance, such legends must have come to prevail under the influence of stories emanating from across India's north-western border. We have already stated that in Persia, towards the middle of the sixth century B.C., a magnificent Empire had come into existence in so dramatic a fashion that it could not have failed to capture people's imagination. We actually see Xenophon presenting the redoubtable Cyrus to the Greeks as the ideal world-ruler; and it is not surprising that for the next two hundred years or more, both in the East and West, his conquests and those of his successors remained the chief topic of political debate and heroic verse. However, India did not have any real experience of a 'paramount' monarchy until some time after Gautama's death. The type of imperial organization, which from the very dawn of history, was being perfected in the home of Osiris and the 'land between the rivers,' was not established in India until a few years after the Macedonian offspring of Zeus-Ammon had breathed his last under a lingering midsummer twilight in Babylon. It was only when the wave of Alexandrian campaigns had finally subsided that Chandra Gupta saw the possibility of, and eventually succeeded in bringing India under a single centralized government. But this great Maurya came about 160 years after Sakya Muni's attainment of Nirvana.

The decay of the old social order based on the twin principles of tribal autonomy and the inviolable sanctity of the family group had set in long before Gautama's days. Irresistible economic pressure and a whole complex of psycho-political forces which those had brought into play, had rendered the tribal-cum-patriarchal republics in India more or less obsolete; their simple economy and social organization belonged to the past rather than to the future. The early tribal institutions and self-governing townships were in advanced stages of disintegration; already, when Gautama was born, several kingdoms had sprung up, and were steadily acquiring power through every conceivable means at their disposal, the methods of expansion ranging from actual wars of conquest to the nobler diplomacy of multiple matrimonial alliances. The monarchical ideal was soon to triumph over the antiquated conceptions which set up family as the primary, and clan as the ultimate limit of social self-consciousness. There was a general tendency among the more influential patricians to usurp the authority of the State whenever an opportunity presented itself. In this manner, there had grown up a conglomeration of petty feudal principalities somewhat on the pattern of the German Duchies or the seven provinces into which England was divided in the time of the heptarchy. These were ruled by autocratic chiefs who found in internecine feuds an admirable outlet for their military ambitions as well as a most exciting pastime. In spite of the rising tide of feudal despotism, however, the old republican institutions had not yet completely disappeared from the political structure of India. Though the Brahmanical literature, with its natural bias in favour of a kingly system of government conducted under direct priestly advice and guidance, maintains a significant silence about these republican bodies, the Buddhist texts reveal that there still survived a number of independent or semi-independent aristocratic republics akin to the city-states flourishing in Greece roughly about the same time. Kapilavastu was one of them.

The exact extent and importance of this republic is difficult to determine, but it is unlikely that it was of any very tremendous account politically or territorially. In both these respects it was probably surpassed by the neighbouring confederacy of the Licchavis, a little way further East. Its independence, too, must have been precarious, hedged in as it was between two expanding kingdoms—Magadha in the south-east and Kosala in the north-west. The patriarchs of Kapilavastu had to play continually for safety between these two rival powers, both greedily looking for a chance to extend their boundaries; and they certainly did not feel very happy about their neutrality. But they appear to have possessed a shrewd diplomatic sense which enabled them to keep the integrity of the republic intact. On the one hand, by adopting an attitude of obsequious amiability towards the kings of Magadha and Kosala, they kept the latter successfully at bay; on the other, whenever occasion presented itself, they did not hesitate to organize punitive expeditions against the weaker hill tribes up north for alleged acts of unfriendliness, and thus replenished the resources of the republic through plunder.

Geographically, the territory of Kapilavastu consisted of a narrow but fertile strip of land 'extending eastward from the point where the Rapti leaves the hills to the little Gandak,' which forms the border between Nepal and India, and is commonly known as the Terai. Above it, like an overshadowing presence, loomed the tantalizing virgin heights of the Himalayas; below on the dim horizon, stretched a seemingly limitless plain with a rich, alluvial soil watered by the sacred stream issuing from the matted locks of Siva, the Dispenser of the Tears of mortals.

The ruling tribe of Kapilavastu was called the Sakyas; and among them in all probability Suddhodana occupied the position of an elected chief with prerogatives corresponding to those of the Roman consul or the Greek archon, but it is not known for how long he held this office. The Sakyas appear to have been a proud people. They traced their lineage to the Sun himself; though all that we are able to gather as to the part He played in the Sakya genealogy is that in some remote and indeterminate epoch he was instrumental in hatching two eggs 'formed from coagulated blood and semen' of their primitive ancestor, also named Gautama. The Sakyas do not seem to have observed very closely the discipline, customs, and usages insisted upon by the Blessed Law-Giver of the 'Aryas.' They were particularly lax in matters bearing on matrimony, and as E. J. Thomas points out, both Gautama and his father married wives of the prohibited degree in flagrant disregard of the 'Aryan' code of Manu. Further, some very highly-placed families of the Sakya clan were suspected of having indulged in incest, which was strictly taboo among the 'twice-born.' The Koliyans, a kindred clan, at one time carried out quite a vigorous campaign of vilification against the Sakyas on this very issue. However, these breaches of 'Aryan' etiquette on the part of their aristocracy did not prevent the Sakyas from proclaiming their solar and 'Aryan' descent from the housetops.

With his characteristic dislike of humbug Gautama was later on to ridicule the whole conception of pride of birth and race. But even his influence did not succeed in sterilizing the 'Aryan' cult which was abroad. The *soi-disant* 'Aryan' conquerors of India, hailing from the wildernesses skirting the northern shores of the Black and the Caspian Seas, were already a most bewildering mixture of ethnic strains—held together by certain linguistic ties and a community of economic interests which found expression in loyalty to a common pastoral Myth-Complex. While forming part of a widespread migratory drift which began in the second millennium B.C., one day they had packed their tents and started out in search of food and fresh pastures towards the Rising Sun. In the course of their long and difficult journey up the valley of the Oxus, across the picturesque plateau of Iran and that enormous reservoir of water—the Pamirs, or Roof of the World—through the narrow defiles of the Hindu Kush which suddenly unfolded before them vistas of a magical land of golden harvests and green abundance, their original, predominantly blond and long-headed stock had acquired an appreciably broader skull and darker complexion. This facial transformation bore eloquent witness not only to fierce battles fought against the elements, but also to their sexual incontinence which had involved them in much lusty and fruitful strife with amazons outside the tribal pale.

The process of miscegenation was naturally accentuated after their settlement in India, in spite of tub-thumping tribal law-givers who laid down severe penalties for exogamous intercourse; and indeed the so-called 'Aryan' invasion of India ended in the gradual absorption of a virile, but numerically small, alien group into the indigenous population rather than in 'Aryanization' as it is usually represented. As time passed, and by slow degrees, the immigrants were drawn into the interior of the Indo-Gangetic basin and the distinction between them and the original inhabitants of the country disappeared for all practical purposes. By the time of the Buddha, ethnic assimilation had reached a stage at which it was impossible to distinguish between mixed breeds and the 'pure' northern strain. And as the Sakyas had penetrated further than most other clans who claimed an 'Aryan' pedigree, it is doubtful whether they had retained any traces of their remote 'noble' origin. If there was any 'Aryan' mystery still left in their blood, it had almost certainly been diluted by large, very hygienic infusions from Mongol and Dravidian sources.

Indo-Germanic enthusiasts have long cherished the image of a blue-eyed and blond Buddha; and Professor Monier Williams even suggests that Gautama was 'a six-footer.' However, the balance of probability points rather in the opposite direction. All views in such matters can at best be intelligent conjectures, but the likelihood of the Buddha being a 'pure Aryan' and a tall blond, is very slight. As far as appearances went, it seems probable that Gautama's features conformed closely to the early figures we find in the Museum at Sarnath, or better still to the type commonly represented in icons from Nepal. The sculptured idealizations of the Graeco-Buddhist school of Gandhara, which possess at once something of the athletic grace of an Apollo or Hermes and the muscular strength of a Meleager, can bear but little likeness to the Tathagata, who was never distinguished for physical strength, and whose earnest and austere exertions in search of the supreme wisdom had left him, as we shall see, with a ruined liver.

Though in Kapilavastu they conducted the affairs of state in a democratic manner, this did not mean that the general social fabric of the Republic was vastly different from those parts of India where autocratic rule had established itself. Indeed, throughout the country, from the swiftly-flowing Indus down to the Bay of Bengal, the social background was fairly uniform. This uniformity may, in a sense, be described as a negative rather than a positive condition: that is to say, everywhere the social organizations were still fluid. Class-divisions had not yet become fully crystallized; society was grouped within an adjustable framework. It is true, immediately after the arrival of tribes from the barbaric West, there had emerged between conquered and conquerors a clear-cut distinction due to the lighter shade of pigmentation among the latter. The Vedic legislators had defined this distinction in the significant term *Varna*—colour. They had used this and the already established precedents to determine the rights of '*connubium* and *commensality*' as an expedient basis for their social classification. But before many centuries had elapsed, the situation had grown far too complex to be dealt with on these simple and arbitrary principles. As we have seen, the constant influx of a large number of the Dravidians and Kolas into the 'Aryan' fold soon deprived the colour bar of whatever efficacy it at first possessed. Since nearly every one showed traces of a darker complexion, in varying shades, it became impossible to maintain the differentiation of society into Blacks and Whites. The concept of *Varna*, it was obvious, had either to go or change its meaning. Eventually, it was retained, but only by giving it a significance totally different from that implied in its original context.

Together with the complexities arising out of racial fusion, the natural division of labour evolved by a society which, though mainly agrarian, had also attained a high level of urban life, resulted in a social stratification never contemplated by the Indo-Germanic cattle-breeders. The association of distinct privileges and possessions with certain functions had led to a very natural wish on the part of those who enjoyed these blessings, to perpetuate and pass them on to their heirs. To this end they sought divine assistance through the mediacy of priests. These latter were willing to oblige by sanctifying the *status quo* on condition that their own prerogatives were made equally secure. The sense of property was increasing. Race-consciousness had long been replaced by class barriers; and where racial feeling existed at all, it was more clearly than ever before a convenient pretext for stabilizing the interests of certain groups which had accidentally come to acquire power and prestige. Thus practically all the elements which go to the making of the rigid demagogy of caste were already evident. But they were still in an embryonic state. The caste as a properly rounded, coherent system had not yet been established. More than a thousand years were to pass before that was fully achieved. The consummation of the tetrachotomous corporate spirit of Hindu society was attained only when the lingering ghost of Buddhism in India had formally and finally been laid.

In this rough, still unfinished social scaffolding it is customary to place the Brahmans, or 'order of philosophers' as the Greeks euphemistically described them, at the topmost rung of the ladder. Actually this does not appear to have been the case. We may seriously doubt if the Brahmans ever enjoyed the supremacy which historians are so fond of attributing to them, except perhaps during the prosperous days of Brahmanical revival in the seventh and eighth centuries of our era. Their lot at the time of the rise of Buddhism seems to have been very far from enviable. It is true that on the strength of their intimate relations with the Godhead, as sole custodians of the secret knowledge which alone could make sacrifice effective and help the dead, they claimed social precedence over all other classes, and in their more inflated moments even regarded themselves as the 'King-Makers.' It may further be granted that the poorer classes among the laity unprotestingly accepted their authority in matters spiritual, and looked upon the Brahmans as influential mediators between them and the Deity, without whose intervention they could not hope to improve their unhappy lot in their next incarnation. But the Brahmans stood on an entirely different footing in relation to the nobles, the clan chiefs, the rich merchants, and growing body of filibustering despots.

This heterogeneous, plutocratic minority formed the really dominant class at the time, and exercised unquestioned power in secular affairs. Its members treated the Brahmanical presumptions with hardly more respect than the feudal aristocracy usually showed towards the minor dignitaries of the Church in mediaeval Europe. They dismissed with the contempt it deserved the Brahman's pretensions of being of a nobler birth than they, the Kshatriyas. They even spoke of him as 'low-born.' The family priest was more often than not considered as a convenient laughing-stock, a source of after-dinner mirth. His talks about the 'One Manifold' resident in the heart of a mustard seed were received with derision rather than intellectual earnestness or reverence; and he usually passed for a mixture of bombastic pedant and amusing, though unconscious, buffoon. The Brahmans, as a class, are painted in the satirical literature of the period in anything but flattering colours. They encountered much adverse, even libellous criticism from the poets and ballad-singers attached to courts or the households of gentlemen of wealth. Even the epics sometimes depict them as libidinous, thoroughly unscrupulous, covetous, unctuous scoundrels, of whom laymen would do well to beware. In a contemporary polemical text—significantly entitled 'the dog'—we read: 'In former times Brahmans approached only a Brahmani; now they go to Brahmanies and Non-Brahmanies alike, while dogs go only to dogs and to no other creatures.' No doubt these pictures were coloured with a great deal of petty spite and jealousy, but it is none the less important to remember that the estimation in which the nobility, and through them a large circle of their underlings, held the Brahmans was not very high.

As for the Brahman's claim that he is born from the mouth of Brahma, the Lord of Creation, Gautama exposed its absurdity with a gentle irony characteristic of him. 'Surely, Vasettha,' he said to his disciple whom the Brahmans had reviled for having joined Gautama's Order, 'the Brahmans have quite forgotten the ancient lore when they say that they are genuine children of Brahma, born of his mouth . . . On the contrary, Brahmanies, the wives of Brahmins, are known to be fertile, are seen to be with child, bringing forth and rearing children. And yet it is these very woman-born Brahmans who say that . . . Brahmans are genuine children of Brahma, born from his mouth, his offspring, his creation, and his heirs. By this they make a travesty of the nature of Brahma.' This view, which Gautama expressed subtly, was held and expressed often quite violently by the majority of the well-to-do Kshatriyas.

And for the most part, the anointed guardian of the word of Brahma acquiesced in the position of undignified subordination into which he was relegated by the rich. He endured the mockery, the pin-pricks, the ill-concealed insults and indignities to which he was subjected—with a broad and cheerful grin. He did not even attempt to defend himself against provocative attacks on his integrity, though he certainly had a case to make. The example in covetousness and greed was set, he might well have pointed out, by those in high places—an example so edifying that the Buddhist poets, because of their congenital incapacity to appreciate the sport of kings, were moved to bitter lamentations and complained: 'The king, although he may have conquered the kingdoms of earth, although he may be ruler of all land this side of the sea even to the ocean's shore, would still, insatiate, covet what is beyond the sea.' In licentiousness there was little to choose between priest and plutocrat; if the hierophants occasionally tried the devices of *Kama Sutra* on some of the lay daughters who came to them in search of fertility, or made cuckolds of gods by seducing virgins dedicated to the deity presiding over the shrine, the nobles helped themselves even much more freely to their poorer neighbours' wives and daughters. In venality the temporal powers were on a level with dispensers of other-worldly benefits; and the State, whether monarchical or republican, encouraged gambling and prostitution to increase its revenue.

However, the Brahman preferred not to hit back. He defended himself, generally speaking, with a philosophic silence. And for excellent reasons. The Order of Philosophers could ill afford to antagonize the rich and influential laity. It would have been very impolitic to threaten their most paying clients with excommunication, since such a threat could have never been carried out; to excommunicate the rich would have meant striking at the very basis of their own prosperity. The well-being of religious bodies depended on the charity of the well-to-do, that of the priestly class on the prosperity of religious bodies. The precious droppings of the rich were like unto manna for the Brahmanical mouth; the anxiety of the nobility to make adequate provisions for their own comfort and the comfort of their dead in the next world brought fuel not merely to the sacrificial fire, but also kept the priest's hearth glowing and his pot boiling. And, above all, the pot had to be kept boiling. For food is the prime necessity of all living things in all the possible worlds of gods, demigods, and men. The sacred texts were unanimous on this issue. They might propagate widely divergent conceptions of the attributes of the Absolute, but they all placed the same emphasis on the life-sustaining property of food, on its paramount significance in the universal order. Brahma Himself, after he had created the 'worlds and the world-guardians,' had with a wise solicitude bethought himself: 'Now let me create food for them.' In the *Aiterya Upanishad* food had been described as the very first material form created after 'He had brooded upon the waters.' The *Maitri Upanishad* had gone even further and suggested that food itself was an incarnation of the Supreme Deity, and that the act of eating, therefore, was like having a supremely blissful communion with 'the Lord of Creation.'

The *Taittiriya Upanishad* had sung of food in still more rhapsodic terms:

> From food, verily, creatures are produced,
> Whatsoever creatures dwell on this earth.
> Moreover by food, in truth, they live.
> Moreover into it they also pass.
> For truly food is of the chief of beings,
> Therefore it is called panacea.
> Verily they obtain all food
> Who worship Brahma as food.
> For truly food is the chief of beings,
> Therefore it is called panacea.
> From food created things are born;
> By food when born, do they grow up.
> It both is eaten and eats things:
> Because of that it is called food.

And the *Upanishads* were the voice of truth. Verily, there was nothing in heaven or earth to compare with the ecstasy of food. Hence, blessed were those who provided the priest with his indispensable panacea. They were more than blessed: they were beyond all blessings, beyond good and evil—like Vishnu, the All-Supporting. Such were the practical cosmogony and metaphysics cherished by the bulk of the Order of Philosophers.

Of course, in this hierarchy, there were a few individuals of great wealth and power. Hinduism had, no doubt, its counterpart to the Christian Lords Spiritual. As beneficiaries from flourishing shrines strategically situated in the holy places frequented by pilgrims seeking absolution, they vied with the Lords Temporal in display of pomp and splendour. Sunk in the soft recesses of well-upholstered priestly thrones, untroubled by thoughts of the morrow, they could afford to worship the Deity in a less gastronomical shape. But not every hierarch was so fortunate. Like the princes of the Church, rich priests formed a small minority. The destitute Brahman, like the poor clergyman, was, and has always been, a more common figure—even in this land of religion.

The Kshatriyas monopolized the earth and the fullness thereof; heaven and its infinite beatitudes were controlled by the priestly Order. But the great multitude had no say either in the earthly or heavenly government. The majority were absorbed in agriculture, handicrafts, and small trades. These tillers of the soil and producers of goods, living their obscure well-disciplined lives in remote villages or on the outskirts of towns, were left very much to themselves so long as they ungrudgingly continued to render unto Caesar what Caesar arbitrarily claimed as his share of their labour. The craftsmen and those engaged in the minor occupations which are the concomitant of urban life seem to have won a certain political status. A skeleton guild system on occupational lines had probably already been evolved. We hear of a number of recognized professions and their guilds. Buddhist literature gives details of about twenty of these guilds, such as, for instance, those of carpenters, masons, workers in metal, weavers, potters, cobblers, tanners, jewellers, carvers in ivory, cooks, confectioners, barbers, shampooers, bath-attendants, painters, garland-makers, basket-makers, workers in rush, fishermen, hunters, trappers, charioteers, elephant-riders, archers, and in later accounts, scribes and accountants. These organizations had rights somewhat similar to those possessed by the guilds in feudal Europe. But the agrarian masses were almost completely ignored by the ruling classes, except in so far as they were a major source of revenue and taxation. In local affairs, it is true, they were allowed a large measure of self-government. Moreover, because of a predominantly communal mode of agriculture, and the almost complete absence of a parasitic landlordism and excessive population pressure on land, they had not been reduced to that state of economic servitude and distress which has been the lot of Indian peasantry in more recent history.

However, in spite of their being comparatively well off, they had little say in the conduct of the State, and were consigned to such a political limbo, that while telling Gautama of the various profitable occupations by which his subjects could enrich themselves, King Ajatasattu of Magadha did not mention agriculture as one of them; and Gautama had politely to remind his royal visitor that he had left by far the largest, and from the point of view of his majesty's treasury, most important section of the community altogether out of his reckoning. Later, when the Greeks came to India on their marauding expedition, their sensitive political conscience seems to have been shocked to see the Indian peasant walking between two armies to his field, utterly refusing to fight, heedless of the great issue at stake. This seeming apathy may really have been due to the shrewdness which made the peasant realize that it would make no difference to his plight whether he had light-skinned or dark-skinned tax-collectors.

Below the social grade represented by the peasantry and handicraftsmen, there was a large nondescript population of menials and pariahs, commonly called *chandalas* and *pukhasas*. For all practical purposes, these were completely outside the social pale; and, as might be expected, they gradually became a single class, united by common disabilities, which to-day we know as the Untouchables, or *Sudras*. Technically at least they could still be considered freemen. But this freedom was mainly illusory; they were free only to hire themselves out as beasts of burden, or do work to which a deep social stigma was attached. Of positive rights either in the secular or spiritual sphere, they had none. Being of 'one birth only' they were outcasts both from human society and divine grace. They could not even claim a fixed habitation because, as the Code of Manu clearly states, 'being always distressed for subsistence' they must live wherever they could hope to gather a few crumbs. They could not aspire to enter the Kingdom of Heaven, or otherwise improve their wretched lot by successive stages of Karmic evolution, for the good reason that they were forbidden to hear, recite, or even memorize the magic mantrams contained in the well-guarded, carefully sealed sacred texts. During the Golden Age of Indian culture under the enlightened Guptas, there is definite evidence that the persecution of the *Sudras* had reached a pitch of hysteria scarcely surpassed in cruelty even to-day. But even before the dawn of this Golden Age to which every 'twice-born' Hindu looks back with pride, Hindu society did not show any mercy in its treatment of the *Sudras*. A comparatively early law prescribes a diabolically ingenious penalty for any *Sudra* caught in the act of infringing the Brahmanical copyright on the Vedas. 'The ears of a *Sudra*,' it lays down, 'who listens, intentionally, when the Veda is being recited, are to be filled with lead. His tongue is to be cut off if he recites it. His body is to be split in twain if he preserves it in his memory.' The *Sudra*, in other words, was deprived even of the right to save his soul.

Yet these outcasts who had 'one birth only' were by no means the lowest of the low. The *Sudras* at least had one birth; but, doubtless, there were others who had no birth at all. This group ranked even lower than they; it consisted chiefly of slaves caught in regular slave-hunting raids, or such persons as had forfeited their legal freedom for having committed a crime. Never having been born, or having lost their birth, logically, they could claim no human rights; they were completely at the mercy of masters who were anything but merciful. Slavery in its glaring Greek and Roman variety probably was not practised in Gautama's India; and we may agree with Rhys Davids, that India at the time had no parallel to conditions prevailing in the Greek mines, the Roman *latifundia*, and the more modern plantations of Christian slave-owners. But that, of course, does not justify the inference that slavery had no place in the social organization of ancient India. The oft-repeated statements of an enthusiastic orientalist like Megasthenes on this point ought to be treated with cautious scepticism. This Plenipotentiary and Envoy Extraordinary of Seleucus Nicator at the court of Patliputtra who found time from his many diplomatic duties to keep a remarkable diary of his stay in India, is certainly very entertaining and informative, but he is very far from being impartial and disinterested. His experience was mostly limited to Court circles; and it is doubtful if he ventured much further than his own Embassy grounds and the gilded threshold of the royal audience-hall. At best he could not have seen much more than what the authorities were anxious to show him. So that his *Indika*, or rather those parts of it which have come down to us in the form of quotations in Strabo and other ancients, can hardly be accepted as giving a true and comprehensive picture of the state of the country. Indeed, his account is in obvious conflict with Buddhist records of roughly the same period. While he would have us believe that 'Indians do not even use aliens as slaves, and much less one of their own countrymen . . .' the Vinaya texts tell a different tale. We learn from these highly outspoken compilations that the Buddhist Brotherhood was forced by circumstances to make stringent regulations against the admission of slaves into the Order. Evidently, it was necessary to take this drastic step, which from a Buddhist point of view is patently uncharitable, because a great number of runaway servants and slaves had flocked into the monasteries to escape the inexorable lash and domestic drudgery, and as a consequence the slave-owning interests had 'got vexed, indignant, and murmured.' Had the Order refused to give them satisfaction, there is no doubt the slave-owners would have sought legal redress. This was the position with regard to slavery some two hundred years after Gautama's death. Conditions could scarcely have been better in his time.

Such was the confused social scene with which Gautama was confronted when he grew up. The situation contained the germ of practically all the class-alignments which have characterized latter-day Hindu society. But social differences were not yet fully defined; the class moulds were still not set. Many centuries had to elapse before they hardened into the water-tight layers of the inviolable caste system.

Love has, of course, always mocked at codes and customs; and though as ever social snobbery was the chief basis of all existing marriage conventions, unions between men and women belonging to widely different stations in society were not unknown in the days of the Buddha. Instances are on record of romantic princelings seeking conjugal bliss in the embrace of bewitching commoners of their fancy; and we hear of ladies of the highest rank leaving their palaces and risking social ostracism for the love of sturdy stable boys. Yet love was not the only motive which induced people to behave in a democratic manner, ignoring class conventions. On the contrary, the truly democratic nature of the period is best illustrated by the fact that in breaking through class barriers men were, generally speaking, moved by practical rather than glamorous considerations. Sprigs of nobility, finding that the vocation of a noble, for all the prestige and honour attached to it, profited them but little in practice, did not hesitate to turn traders in the hope of making large fortunes and otherwise improving their worldly prospects. Brahmans were persuaded by reasons of their own to abandon their hereditary life of solemn routine and take up some healthier and more invigorating profession such as, for example, that of a soldier, or hunter, or trapper, or even wheelwright. Thus, although the social boundaries had already been marked out, they were rather flexible and tended to merge into one another, especially at the edges. For the rest, as among all the peoples of the earth, and in all times, men were prepared to go to any length of buffoonery and meanness, suffer any hardship and humiliation, and commit any crime and cruelty in order to make sure of their panacea, familiarly known as 'daily bread.'

The Sakya republic had a population of about a million; and Kapilavastu itself was the capital of the republic. There were a number of other minor townships in the realm, the best known of which were Samagama, Silavati, and Devadaha. The last-named of these has a certain historical importance, being the home of Gautama's mother. Otherwise, however, Kapilavastu was the only township of a mentionable size in the Sakya territory.

It is now almost impossible to establish its exact situation. The town was sacked and destroyed by Vidudabha, successor to King Pasenadi of Kosala, three years before Gautama's own release from the tentacles of birth and death. This infamous son of a famous and enlightened father, who usurped his father's sceptre with the aid of an unscrupulous general in Pasenadi's army, cherished a very deep grievance against the Sakyas and availed himself of the first opportunity to avenge himself upon them. He perpetrated a wholesale slaughter of the Sakyas, sparing none that he could lay hand on—not even sucklings. The destruction of Kapilavastu was carried out by him with such a painstaking thoroughness that few traces of it were left for posterity to admire. The celebrated Chinese pilgrims when they visited India in the fifth and seventh centuries were unable to locate the position of the city with any degree of accuracy. Fa Hsian (A.D. 399-414), in the account of his pilgrimage to the Buddhist holy land, remarks, not altogether without a keen sense of disillusionment, that the place which was pointed out to him as being the site of Kapilavastu, where the Buddha had spent his youthful years dallying in love and luxury, was but a dismal tumble-down village. 'No king nor people are to be found,' laments the Celestial who was probably hoping to spend many exciting days exploring the remains of palaces, pagodas, and pavilions, and all the painted pomp described in the story of 'the Great King of Glory' which had achieved a very great popularity in his times. 'It is,' he adds ruefully, 'just like a wilderness, except for priests and some tens of families.'

Hsuan-tsang (A.D. 629-645), who followed in the footsteps of the Master two hundred years later, had to face a similar disappointment at being taken round an insignificant deserted ruin. But the place he was shown was probably not the same as that visited by Fa Hsian. Already, apparently, the people of the neighbourhood were anxious to attract lay tourists, sightseers, and pilgrims to their own particular localities, and were making rival claims as to the place of the Tathagata's nativity.

The discrepancies which crept into the tradition at a very early date have never been resolved. All that may now be said on the point is that the Sakya capital was situated on the banks of the river Rohini, a small tributary of the Ganges, about one hundred and twenty miles due north of Benares, which at that time and for almost two millenniums to come was indisputably the premier city of India, famous seat of Brahmanical learning, and religious centre of the whole Hindu world. It is not at all unlikely that Kapilavastu lay at the junction of two important trade routes: the one linking Benares with Savatthi, the capital of the principality of Kosala, and the other Savatthi with Rajagaha, the capital of the growing kingdom of Magadha. If so, this must have contributed in no small degree to the prosperity of the Sakya republic in which Gautama's father, as the first citizen of the realm, was an interested party.

However, even at the height of its glory, Kapilavastu was not very remarkable for its magnificence and splendour considered as a city. Very late in life, when Gautama was staying at Kusinara, a small township in the territory of the Mallas, and awaiting the end which he knew to be fast approaching, we hear Ananda saying to him with characteristic naïvety: 'Let not the Blessed One die in this little wattle and daub town, in this town in the midst of the jungle, in this branch township. For, Lord, there are other great cities, such as Kampa, Rajagaha, Savatthi, Saketa,

Kosambi, and Benares. Let the Blessed One die in one of them. There there are many wealthy nobles and Brahmans and heads of houses, believers in the Tathagata, who will pay due honour to the remains of the Tathagata.' Evidently Ananda did not regard Kapilavastu, the Blessed One's native town, among the places where it would be proper for him to attain Nirvana. And quite rightly so. For at best Kapilavastu was no more than a second-rate, though fairly flourishing, provincial town. It was built without much attention to the requirements of town-planning, architectural layout, or sanitation. There were, of course, a few rich patricians and merchants. They had their stately homes with imposing entrances; their stone exteriors carved into delicate patterns of wreaths, creepers, ribbons, and dragon's teeth; their vast halls and pavilions which were so profusely covered with frescoes that they were known as picture-galleries; their labyrinthine inner quarters reserved for the exclusive use of their womenfolk; and their large, closed-in gardens where these favoured ones of fortune could frolic with their concubines in ease and privacy. Some of these palatial mansions were several storeys high, and we also hear of a peculiar type of building constructed on a pyramidal design vaguely reminiscent of the Babylonian Ziggurat. But there were not many of these architectural marvels. For the most part the streets were narrow, congested, and badly, if at all, paved; the houses dingy, unventilated, ramshackle constructions of wood, bricks, or roughly hewn stones, lacking in the most rudimentary sanitary arrangements; the drainage system conspicuous by its absence. But in all these misfortunes, it should be added, Kapilavastu was by no means unique; it merely shared certain features common to practically all ancient and mediaeval Indian cities. This disregard of the elementary demands of social hygiene in planning towns and building houses had serious consequences for the general level of health among the town-dwellers, an appalling percentage of whom were afflicted with skin-diseases of all kinds, ranging from leprosy to ordinary rashes. Even the highest strata of society were not altogether immune from these dreadful visitations; there were frequent outbreaks of small-pox, boils, and local sores among the Court circles and the royal seraglios.

The best and most prosperous parts of the city were situated within strongly buttressed ramparts; and as a further precaution against attacks and raids, a double moat of mud and water engirdled the city. The times were peaceful, but not too peaceful. The outlying suburbs were inhabited mainly by outcasts, menials, and other social refugees who had no political rights and few possessions. Their dwelling-places were huts of mud and thatch which were little better than dunghills. Inside the city, the most important public building was the Moat-Hall, or the municipal town hall where the citizens, young and old alike, gathered to discuss affairs of state and decide issues of public importance. Probably, it also served as the law courts, where the elders of the city met periodically to dispose of criminal and judicial cases.

However, neither politics nor the dispensation of justice, except on rare and sensational occasions, were of any very great interest to the ordinary citizens. Indeed, they were glad to leave the conduct of government to their betters and to be left in peace to pursue their proper occupations. For excitement, there were all sorts of amusements available. There were the temples where the contrite and conscientious congregated daily at fixed hours to pray and offer sacrifices for the redemption and elevation of their souls. Gamblers flocked into the public gambling hall, and in the oracular rattle of the dice, and the tense expectancy of the game, found a satisfying substitute for the Good, the True, and the Beautiful. Then there were the opportunities for venal love, with its promise of pulsating bliss for those whose inclinations were of a less abstract nature. In the evenings, usually after a much-needed hygienic treatment at the hot-baths, mystics desirous of curing the ills of the soul through ministrations of the body made their way to the quarters inhabited by the priestesses of Venus. There, in the frenzied friction of flesh against flesh, the separate entities of a four-dimensional universe struggled to cast off the burden of their solitude; forgetting in their momentary abandonment the distinctions between *Prakriti* and *Purusha*, between Being and Non-Being, between the Causal and the Accidental, between the Eternal and the Transient.

III

SOME EARLY EVENTS

ZOROASTER, according to an ancient tradition recorded by Pliny, astounded every one around him by breaking forth into hilarious laughter the very day he was born. The Buddhist legend attributes something much more startling to Gautama. As soon as he emerged from his mother's womb, the Bodhisattva 'examined the four quarters, the intermediate quarters, the zenith and the nadir, ten quarters, and not seeing any one like himself he said, "This is the northern quarter," and took seven steps. While Mahabrahma held a white parasol over him, and Suyama a fan, and other divinities followed with other symbols of royalty in their hands, he stopped at the seventh step, and raising his lordly voice, "I am the chief in the world," he roared his lion-roar.'

The reason for this strange behaviour on the part of the infant Buddha remains obscure, though it is possible, as E. J. Thomas observes, that the account involves certain delicate doctrinal aspects of the incarnation of a Bodhisattva or a potential Buddha as it was envisaged by the Buddhist mystagogues. However, there is no need for us to enter into a discussion of these superfluous esoteric subtleties. The phrase 'lion-roar' itself is not quite so disturbing as it might appear on the face of it; if it recurs with a monotonous frequency throughout the sacred texts of the gentle and peace-loving Buddhists, it implies no innate aggressiveness, but is probably just a figure of speech.

The whole Discourse of 'the Wondrous and Marvellous Events' is so utterly incredible that we are tempted to suspect it of being entirely bogus. There is not a single detail in the surprising incidents which heralded Gautama's birth to convince us that he was an infant prodigy who began to turn the Wheel of Law and propound 'the Doctrine, good in the beginning, good in the middle, good in the end, complete in the letter and the spirit, whole, pure . . .' while still in the cradle. On the contrary, we may reasonably suppose that in his childhood he behaved very much like an ordinary child. Even during his early adolescence, it seems, if he was remarkable at all, it was for a certain spirit of intelligent indolence which made him neglect 'the manly exercises' fashionable among the budding youths of noble birth. His indolence even aroused some unkind comment. In this, he was a complete contrast to his cousin from the maternal side, Devadatta, whom the legend features as a life-long rival to Gautama, and who later on joined the Order, but apparently for no other purpose than making himself a nuisance to the Buddha.

Devadatta is a personage of some importance in the Buddhist story. It has sometimes been suggested that in him we have a close parallel to Judas Iscariot of the Gospel drama. The comparison is rather far-fetched, and if there is any kinship between Judas and Devadatta, it does not go beyond superficial resemblances. The picture of Devadatta which we get from the Canon lacks both the depth and complexity of the Biblical study in chiaroscuro. Judas, as Middleton Murry remarks, was probably the only person among the twelve Apostles who understood the significance of Jesus and had a measure of his suffering. His betrayal of the Man of Sorrow was an act of supreme love; and it was perhaps quite appropriate that the act should have been consummated after a kiss.

For the realization of so subtle a character, a tragedy of passionate love is essential; and the Buddhist legend, in none of its variations, ever attains the intensity of a Passion Play. Although a Mongol work belonging to the Mahayana, or the Greater Vehicle, suggests that 'the sublime Bodhisattva Devadatta during five hundred births, in which Buddha was going through the career of a Bodhisattva, inflicted on him all possible evil and suffering . . . in order to establish the excellence and high qualities of the Bodhisattva,' the explanation sounds a little hollow. No convincing argument has ever been forthcoming to show that the recurrent inconveniences which Devadatta caused the Bodhisattva were prompted by any other motive than petty malice and childish rancour. The general impression received both from the Canon and the Chronicles is that Devadatta's was the jealous nature of a mediocrity, anxious to get on in the world at all costs. From very early youth we find him engaged in spreading all sorts of malicious reports about Gautama, making it generally known among the Sakya clan that Suddhodana's offspring was an indolent good-for-nothing wastrel quite unworthy of his noble birth, and even casting serious aspersions on his manliness. The legend goes to the extent of arranging an open tournament, and tries to make out that the initiative for it came from Gautama himself who wished to confound his enemies by displaying his skill in 'the twelve arts' and defeating them at their own game. At the public bout which duly takes place on an appointed day, Devadatta together with a number of other noble youths contest the laurels with Gautama, but the latter eventually carries the day by accomplishing the Herculean feat of whirling a dead elephant by its tail and hurling it two miles over the ramparts. The story is far too fantastic to be true; nevertheless, it indicates the relation in which the two cousins stood to each other from the earliest period in their careers.

The rivalry seems to have been almost entirely on Devadatta's side. Gautama was not particularly interested in getting on or making good in the world; his was not the temperament of a go-getter. It is doubtful if he could have ever been persuaded to compete for honours with his cousin in any sphere, or any one else; for one thing, even in his early youth, he seems to have been far too self-conscious to indulge in exhibitionism. There could be no better proof of his complete disregard of considerations of worldly loss and gain than that he allowed Devadatta to become a member of the Order, though he could hardly have had any illusions as to the intentions which had induced his cousin to accept the Buddhist faith. It is still more significant of his good-will towards Devadatta that he allowed him great latitude, though from the very day of his entry into the Buddhist fold Devadatta concentrated all his energies on the single aim of bringing about a disruption of the Brotherhood. The canonical accounts contain references to his insidious efforts to create a schism in the Order and bring Gautama into disrepute by every possible means. The later commentaries improve vastly on the canonical stories of Devadatta's infantile schemes of treachery: they attribute to him heinous deeds of perfidy, including three successive attempts on the Blessed One's life, the murder of a famous nun named Uppala-Vanna, and worst of all, adultery with Gautama's wife (whom the account in question represents as being merely a distant cousin and not, as in the Pali Chronicles, a sister of Devadatta) after Gautama's Renunciation.

These gruesome tales of crime and treachery are probably nothing more than fanciful inventions of the latter-day romanticizers who must have felt that the legend was incomplete without a proper 'villain-of-the-piece,' and found in Devadatta a character which could easily be twisted into the most sinister interpretations. But that there is a historical basis to some of the stories concerning Devadatta seems fairly certain. It is extremely probable, for example, that he was a participant, if not actually the moving spirit, in the conspiracy of Ajatasattu, the Crown Prince of Magadha, to depose his father Bimbisara, and then starve him to death. Devadatta was interested in having that kindly, if dissolute old man safely out of the way, because he happened to be a personal friend and benefactor of Gautama.

However, in spite of constant provocation from Devadatta, Gautama does not appear to have taken any steps to safeguard himself against this scheming genius. There is only one important action on Gautama's side which is recorded: and it had to be taken because Devadatta had forced the issue by founding a separate Order of his own. This left Gautama no other choice than formally to decree that: 'in future, whatever he may do or say, Devadatta shall be considered as acting or speaking on his own behalf,' and not as an accredited spokesman of the Brotherhood. It may be pointed out that even this Act of Proclamation, did not speak of a formal expulsion from the Order. Indeed, the consideration shown by Gautama in his treatment of his exasperating cousin is quite surprising. Partly, his forgiving gestures were the outcome of his truly generous nature; partly, they sprang from the intellectual conviction at which he appears to have arrived rather early in life, that 'not by hatred is hatred appeased.'

Yet his tolerance and reasonableness did not improve the situation; if anything, it served further to intensify Devadatta's hatred and dislike of him. His one great grievance against Gautama might very well have been the consideration with which the latter treated him. Paradoxical as it may sound, this was a very real and genuine grievance; a grievance sufficiently acute to disturb the equanimity of a far more balanced person than Devadatta. It is in no way surprising that it drove Devadatta to the verge of madness, and in his desire to avenge himself upon Gautama for an imaginary wrong, he forgot even his own interests and frequently made a perfect fool of himself.

The name-giving ceremony of Gautama took place on the fifth day after birth. His mother was perhaps still writhing on her death-bed, but not a word is said about her. Evidently, it was not considered of any importance. There are, on the other hand, detailed and highly coloured accounts of portents and prophecies as to the miracles which the infant Bodhisattva was going to work on coming of age. As was customary among the nobility, the ceremony was made an occasion for the distribution of free meals among a large number of Brahmans. According to one account, a hundred and eight priests actually had the privilege of being invited to the house of the President of the Sakya Republic to be fed. Of these worthies, eight were supposed to be initiates, well versed in the art of soothsaying and interpreting bodily marks, like sage Asita who appeared on the scene immediately after the Bodhisattva's birth. They confessed to having had a dream on the night of Gautama's conception, in which the future of the child had been revealed to them. Nevertheless, seven of them judiciously tempered their prognostications with a saving element of ambiguity, and raised two fingers in a somewhat equivocal prophetic gesture. They were not prepared to commit themselves with any finality, but cautiously prophesied that Gautama would achieve fame either in the vocation of a great monarch or as a Buddha. The eighth, however, a young fellow known as Kondanna, left no uncertainty in the terms of his announcement. Being a better informed and a bolder spirit than his colleagues, he lifted a solitary finger in solemn warning, implying that there was no room for doubt as to the shape of things to come: the baby was destined to grow up to be the Light of the World. He also announced that Gautama would be urged to renounce his worldly state by four signs—an old man, a sick man, a corpse, and an ascetic; and that if Suddhodana wished his son and heir to concentrate his mind on improving family fortunes in heaven and on earth, instead of foolishly attempting to show the Perfect Way to a misguided humanity, he would be well advised to guard against the possibility of the boy seeing the four ominous sights. This stylized form of prophecy is plainly a happy after-thought. At the same time, there is no reason why, after a hearty meal, these prophets should not have been inclined to take an extremely optimistic view of Gautama's future and suggest that so far as he was concerned destiny had arranged everything for the best in the best of all possible worlds.

The boy was named Siddhartha, which literally means 'he who has accomplished his purpose.' But there is nothing to show that this particular name was decided upon in order to hint at coming events; and, as Rhys Davids remarks: 'other Siddharthas are mentioned who were not at all peculiarly successful in accomplishing their desires.' The habit of giving grand names is universal. It is quite possible that the name Siddhartha, like so many other grandiloquent epithets which are employed for Gautama, is merely a title of latter-day origin, rather than a name. One may agree with Rhys Davids when he says: 'These expressions, like the "Swan of Avon," may have had very real significance in moments of poetic fire. But their constant use among the Buddhists tended, not to bring into clearer vision, but to veil the personality of Gautama, and their constant use as names by modern writers arises simply from mistake.' And though 'to the pious Buddhist it seems irreverent to speak of Gautama by his mere ordinary and human name,' there is hardly any doubt that it was by this name he was known not only to his contemporaries, but also for a considerable time after his death.

With the death of Gautama's mother, the responsibility of nourishing and looking after him fell upon his aunt, Suddhodana's second wife. Unfortunately, we cannot gather very much about this self-effacing lady from any of the available sources. She remains entirely outside the legendary limelight, but it may be assumed that she attended to Gautama's needs with tender care and devotion, especially as he happened to be the only child of his father, with all the parental hopes and ambitions centred round him.

To say anything about the relation which existed between Gautama and his foster-mother, we have to rely largely on conjecture. But it seems unlikely that a mother-and-son tie was ever formed between them. This much may be surmised from indirect evidence. For the death of Gautama's mother at the time of his birth appears vaguely to account for a void not only in his personal life, but also in the subsequent growth of Buddhism as a world-religion. It is a curiously revealing fact that Buddhism has always remained lacking in a proper Madonna cult. Only in its Far-Eastern variety do we encounter anything approaching a full-bosomed, broad-hipped Madonna; and it is obvious that she has been introduced into its otherwise spotlessly chaste, doctrinal folds in a somewhat clandestine manner, with the intention of luring into the ranks of the faithful those numerous desolate souls who never outgrow their early passion for the sweet juice of the maternal breast, and go through life perpetually languishing for want of it. But even there, the charming and solicitous Kuan-Yin, or the Goddess of Mercy, remains a deity of secondary order.

Unlike the Holy Virgin, she does not form one of the chief pivots of worship and adoration. It is not altogether fantastic to suggest that this absence is in some way, no matter how remote, connected with a corresponding absence in the life of the founder of the religion.

On Gautama's own development, the incident may have exercised an influence, which because of its subtlety has so far never been suspected. Probably it was a purely negative influence. But a negative factor can prove more decisive, in some respects, than a positive. In moulding an individual's character the influence of an absence can sometimes be much more important than that of a presence. Thus the absence of a mother-fixation in the background of Gautama's experience seems to explain one of the most striking traits in his character. In a study of his personality one is impressed by the fact that throughout his long life he never formed any intimate relationships. His work brought him in contact with all sorts of people, kings and cobblers, and though he was friendly towards all, he 'allowed intimacy to none.' In his relations with his family, before his Renunciation, there is no suggestion of a sentimental attachment. After his Renunciation this characteristic becomes even more marked. He treated his disciples with a consideration and kindness such as has rarely been shown by any man, and yet even his most devoted companions knew him as a friendly stranger rather than as an intimate friend. This is the more surprising because Gautama was endowed with exceptional depths of understanding, compassion, and tenderness.

There are, of course, a number of reasons to account for the detachment which he cultivated. Partly, it was no doubt the fruit of a long and careful discipline of the mind which he considered absolutely essential to a reasonable way of life. Further, the fact that he was intellectually far above most of the people in his immediate circle, that he had thoughts and feelings far beyond their understanding, must have created an impregnable barrier between him and his companions, making it impossible for him to be communicative except in impersonal matters. However, the absence in his experience of those early emotional ties, which become often the basic pattern for all profound relationships developed later in life, is a factor which cannot altogether be ignored.

Gautama's father was a fairly well-to-do noble according to the actuarial standards of his times, but he was very far from being a millionaire. The stories of his fabulous wealth should be treated with the same cautious scepticism as the lavish accounts of the legendary treasures of the Great King of Glory. Certainly, the passages describing the luxuries with which Suddhodana surrounded his son and heir during his infancy and youth never attain the richly sensuous beauty of the *Mahasudassana Sutta*. There is nothing in them to rival the description of the 'royal city of Kusavati' with its magnificent 'Palace of Righteousness,' built of bricks 'of gold, of silver, of beryl, and of crystal'; its pillared gates and pleasure grounds; and its magical avenues of palm-trees, which, when shaken by the wind, produced a sound so 'sweet and pleasant, and charming, and intoxicating' that gamblers and drunkards would dance together to its music. But even though, comparatively speaking, the account in the *Anguttara Nikaya* may appear to err on the side of understatement, there is no dearth of purple patches. It makes Gautama dwell on the splendour of the palaces in which he spent his early youth in a ruefully nostalgic retrospect:

'I was delicate, O monks, excessively delicate. In my father's dwelling lotus pools had been made, in one blue lotuses, in another red, in another white, all for my sake. I used no sandal-wood that was not of Benares, my dress was of Benares cloth, my tunic, my under-robe, and cloak. Night and day a white parasol was held over me so that I should not be touched by cold or heat, by dust or weeds or dew.

'I had three palaces, one for the cold season, one for the hot, and one for the season of rains. Through the four rainy months, in the palace for the rainy season, entertained by female minstrels I did not come down from the palace; and as in dwellings of others, food from the husks of rice is given to the slaves and workmen together with sour gruel, so in my father's dwelling rice and meat was given to the slaves and workmen.'

Asvaghosha in his sentimental romance based on Gautama's life—a work that has exercised a very profound influence on all subsequent biographical literature dealing with Gautama, and which, despite its tiresome exaggerations, is not altogether devoid of a certain psychological veracity—provides from his fertile imagination a few more intimate details about the Bodhisattva's early life. 'Then he spent his time in those royal apartments,' the prolific contemporary and spiritual mentor of the famous Kushan king, Kanishka, writes in his characteristically euphuistic style in *The Buddha-Karita*, 'furnished with the delights proper for every season, gaily decorated like heavenly chariots upon the earth, and bright like the clouds of autumn, amidst the splendid musical concerts of singing women. With the softly-sounding tambourines beaten by the tips of the women's hands, and ornamented with golden rims, and with dances which were like the dances of the heavenly nymphs, that palace shone like Mount Kailasa. There the women delighted him with their voices, their beautiful pearl-garlands, their playful intoxication, their sweet laughter, and their stolen glances concealed by their brows.'

These and many other similarly fanciful accounts are obviously the result of an incorrigible tendency among the world-renouncers of all ages to dilate upon the things of the world with a nostalgic, almost tortured amorousness. They cannot be accepted as an accurate statement of Gautama's family resources. Suddhodana, though he no doubt loved his son and heir, was not sufficiently rich to build three different palaces for his use and to furnish them 'with the delights proper for every season.' Asvaghosha does, of course, try to give his story a touch of reality by telling us beforehand that from the time of the birth of Gautama, his father increased 'day by day in wealth, elephants, horses, and friends as a river increases with its influx of waters.' But two inventions can hardly make one truth.

The 'pleasure domes' decreed by Suddhodana 'for the sake of ensuring his son's prosperity' and warding off the evil 'destiny which had been predicted of him,' were never built except in the spacious imagination of the chroniclers. But he did provide the boy with all reasonable comforts and conveniences within his means, never grudging him such luxuries as were meet and proper to his station in life. There is little doubt that Gautama's early years were spent in comparative ease and indolence, and that he was not troubled by any thoughts of the morrow.

Of how and where he received his education scarcely anything is known. India in his days possessed quite a number of good universities which trained young men in arts and sciences. The Academy of Takhasila (or Taxila, as the Greeks called it) in the extreme north-west had already achieved a renown which extended from the banks of the Indus to the mouth of the Ganges, and beyond. It had a highly learned band of teachers; kings and nobles sent their sons to Takhasila to be instructed in public affairs and politics, a fact which may be taken to indicate that, though the emphasis was on a scholastic and quasi-religious kind of teaching, humanities were not completely ignored. It was at Takhasila that king Pasenadi of Kosala had been educated, and Givaka Komarabhakha, the Physician and Surgeon Royal to the king of Magadha, acquired his remarkable skill in medicine and surgery. However, there is no mention of Gautama having been sent either to Takhasila or to any of the other famous seats of learning. The *Lalita-vistara* does speak of his visit to a writing-school, but since in the sixth century B.C. alphabet was not yet in common usage and writing as such largely unknown in the country, the story may be rejected as a complete fabrication. The education of Gautama was probably entrusted to a private tutor who, according to the custom, gave him oral instructions in the sacred lore and physical training in the 'manly' arts of archery, wrestling, riding, hunting, etc.

There is no evidence that Gautama ever distinguished himself in any of these pursuits; in fact, the likelihood is that he never evinced great enthusiasm for them. Very little is recorded of his achievements either in the 'academic' or 'manly' side of his studies; and what is recorded is too fantastically spectacular to be credible. Of course, Asvaghosha is anxious to make us believe that 'the prince gradually grew in all due perfection.' But against this, we have the testimony of the southern school of biographers of 'the prince.' They tell a very different story. They unanimously insist that the Bodhisattva was somewhat neglectful of his studies. The matter even became something of a minor public scandal; his relations, jealous of the family's reputation, were forced to complain 'in a body' to Suddhodana of his son's laziness, and impress upon him that such dissipation ill befitted the heir of the elected head of an aristocratic republic in difficult times. On the whole, the southern accounts ring more true than the classical sources. It is very probable that at a fairly early period in his life Gautama began to show that reflective and serious turn of mind which became increasingly marked as the years passed. A personality like his, which had all the essentials of intellect, though not its superficialities, would almost certainly have invited the censure of the unimaginative poor in spirit.

The ancients cherished their own notions about marriage and the proper time for embarking on marital adventure; and these were in some ways radically different from those approved by the self-righteous champions of puritan morality and the upholders of the legal Age of Consent. The ancients, literally believing that all marriages are duly made and consummated in heaven, saw no sense in depriving their children of their divine right any longer than was absolutely necessary, and thus spared them what Remy de Gourmont justly describes as 'a painful transition,' a state that is liable to become 'a torment as soon as it is prolonged.' Throughout the ancient world, early marriages were extremely popular; indeed even in Europe, until the closing years of the eighteenth century, the practice of early marriage was fairly common, and was considered entirely proper and respectable. However obnoxious and sinister the custom of pairing off boys and girls at the first available opportunity after their attainment of puberty, and sometimes even before, may appear to-day, the ancients saw in it an institution which offered distinct advantages to all the parties concerned. As Gourmont in his excellent admonitions in *Le Chemin de Velours* takes some pains to point out, by this means, on the one hand, woman was saved from having to waste 'the third of her sexual life and some of the years best fitted for love'; on the other, 'the parents were glad to be freed from their responsibility, and the husbands, without any illusions about the future, married young girls to be certain of at least one or two legitimate children.'

In India, these practical reasons which urged the parents to seek bed-partners for their offsprings at an early age, were further fortified by religious considerations of a very serious nature. Sages of the highest eminence, from Vasistha down to Yajnavalkya, had laid down the law on this issue in explicit terms: 'He that doth not give his grown-up fair daughter to a worthy wooer,' they had declared, 'let him be held a Brahman murderer. Each time a ripe unwedded maiden has her courses, her parents or guardians are guilty of the heinous crime of slaying the embryo.' And endorsing this dictum, Paraçara, another celebrated Hindu Law-Giver, who held that 'a girl of ten becomes a maid and with this physiologically a perfect woman,' had gone a step further by observing that 'if a girl has reached her twelfth year and has not been given away, then her forefathers in the other world are for ever drinking the blood she sheds every month.'

In view of these strict injunctions and the prevailing opinion of the age, it was but natural that for Gautama there should have been no 'painful transition' between playing marbles and marriage. Tradition maintains that he was married when he was sixteen; for once tradition may unquestionably be accepted. In fact, it is recorded that even before he was properly married, his father had, very sensibly, arranged for his initiation into the intricate mysteries of *Ars Amandi*. Asvaghosha writes: 'The prudent kings of the earth, who wish to guard their prosperity, watch over their sons in the world.' Apparently, Suddhodana had a different conception of 'prudence' and safeguarding the family fortunes. For Asvaghosha adds: 'But this king [meaning Gautama's father], though loving religion, kept his son from religion and set him free towards all objects of pleasure.' Other accounts fill in the details about those 'objects of pleasure,' lovingly dwelling on the amorous escapades of the Bodhisattva. Some of them put the number of the dancing girls in his entourage at eighty thousand, others at forty thousand, but there are also some humbler and more reasonable estimates. These astronomical figures, of course, are not to be taken as having any relation to reality. They may possess some abstruse mystical significance, but for the most part they are mere instances of the oriental tendency to exaggerate.

The genealogy outlined in the Chronicles gives the name of Gautama's consort as Bhaddakaccana, and represents her as the daughter of Suppabuddha, the brother-in-law of Suddhodana. Thus she would be a sister of Devadatta, and cousin of Gautama himself. Other sources give different names, such as Gopa, Bimba, Yasodhara. But, after all, what is there in a name? Bimba or Bhaddakaccana, Gopa, or Yasodhara, the passage of time has bedimmed their separate identities, if they ever had any, into a single abstract note symbolic of conjugal felicity.

Suddhodana's task in arranging a suitable match for his son does not appear to have been by any means easy. Indeed, it is related that he had to contend with considerable difficulties in finding 'a ripe unwedded maiden' for Gautama. If the version of the Chronicles is accepted, it is possible to imagine how the trouble arose. It is likely that Devadatta, who is supposed to have quarrelled with his cousin even over such trifles as a certain goose he had shot, viewed the project of a union between Gautama and his sister with profound hostility. He probably tried his best to dissuade his people from giving the hand of their 'fair and full-grown' daughter to Gautama, and doubtless advanced very good reasons why he did not approve of that particular marriage. It was altogether a weighty matter. Bhaddakaccana's parents, obviously, had no wish to be 'Brahman murderers.' The Hindu law-givers were emphatic that an unconsummated marriage was no marriage at all; they insisted upon 'a worthy wooer.' Wedlock was wedlock only when the husband was capable of his wife, and the wife of her husband. Gautama was 'fair and fit' to look at. But appearances might easily be deceptive. Could Suddhodana's son offer Bhaddakaccana that essential 'support' without which matrimony must remain a mockery, a hollow caricature of its purpose? Evidently, Devadatta did not think so; and he seemed to know. On the other hand, he might be wrong. Apart from the question of causing needless offence to Suddhodana, it was not every day one got an opportunity of finding so eminently suitable a husband for one's daughter—the son of a noble of the highest standing, heir to the first citizen of a flourishing township. There was doubt. Suppabuddha was puzzled. He was more than puzzled: he was faced with a terrible dilemma.

Some of the accounts resolve the issue on the usual dramatic lines. They describe how Gautama, when told of the rumours that were abroad, publicly proved that he was not merely 'fit and fair' to look at, but could 'support' a wife as well as anybody else. He is then shown stringing 'the bow requiring the strength of a thousand men,' striking the string till its sound—which was like that of 'thunder'—roused the whole city, and performing a number of other feats of superhuman strength in a truly 'Aryan' fashion. But the story of this mimicry of nuptial rites, which is clearly derived from the Epics, does not ring true. If any doubts did actually arise in Suppabuddha's mind as to the capabilities of his would-be son-in-law, then he would have made friendly enquiries through diplomatic channels. Eventually, it seems, he was assured that Gautama was in every way capable of his daughter. The marriage took place with due ceremony, though strangely enough, the legend is silent about it. Subsequent events proved that Suppabuddha's apprehensions were entirely groundless: in the fullness of time Gautama's wife bore him a son.

This would also dispose of the legend already mentioned which attributes a 'sheathed member' to the Buddha as being among the thirty-two bodily marks of greatness.

BODHISATTVA OF THE BLUE LOTUS
(Ajanta—*The Director of Archaeology, Hyderabad*)

IV

THE PURSUIT OF PLEASURE

Vivre? Nos serviteurs feront cela pour nous!
Villiers de L'Isle-Adam

DANDYISM is a universal phenomenon. Its character is the same everywhere. It makes little difference whether we are dealing with the ancient East or the modern West, Imperial Rome or Paris under the Third Republic, the Court Celestial of Sung monarchs in the Land of the Middle Earth, or the early dynastic glory in the home of Osiris, Periclean Greece or the Age of Reason in France. Dandies all over the world, and of all ages, are united in a kind of spiritual fraternity: theirs is the oldest freemasonry. There were dandies in Gautama's days who would have been looked upon with absolute approval by the author of *Axel:* they, too, 'lived' by proxy.

'*Le Dandy*,' writes Baudelaire, '*doit aspirer à être sublime sans interruption. Il doit vivre et dormir devant un miroir.*' This is, of course, a rather exacting standard, the achievement of which demands an endurance far beyond human strength. No dandy has yet been able to keep on the sublime level '*sans interruption*,' or to live and sleep always before a mirror. Even the most meticulous among their tribe are apt to have moments of self-forgetfulness. Nevertheless, dandies among Gautama's contemporaries spent a good deal of their time in front of their mirrors. When Megasthenes visited India two hundred years after Gautama's death, he was struck with the care which the Indian dandies bestowed on their looks, and the way in which they tried to dress uncouth nature to advantage. 'In contrast to the general simplicity of their style,' he comments admiringly, 'they love finery and ornaments. Their robes are worked in gold, adorned with precious stones, and they wear flowered garments of the finest muslin. Attendants walking behind hold umbrellas over them; for they have a high regard for beauty, and avail themselves of every device to improve their looks.'

Other sources throw light on the happy cultural and economic issue of their 'high regard for beauty.' As a result of their vanity about their persons, and their praiseworthy anxiety to appear sublime at all costs, the luxury-trades, whose prosperity forms the best index to the cultural refinement achieved in any age, were the most profitable in India at the time, and craftsmanship in the manufacture of all kinds of luxury-articles had attained a very high degree of artistry. In particular, we are told, the guilds of tailors and barbers flourished as they had never flourished before. The barbers especially did a roaring trade; theirs was the most ambitious of all the crafts directly or indirectly concerned in flattering the personal vanity of mankind. While the sartorial skill of the tailor, for instance, stopped short at dressing Nature to advantage merely from the outside, the barbers probed deeper into reality. They went straight to the thing itself; handled Nature in the raw in order to fashion it into a 'sublime,' or at least more human shape. As elsewhere, they were recognized specialists in the difficult operations of hairdressing, arranging of coiffures, shaving, trimming of beards, shampooing, and manicuring. But apart from these well-known processes, which have long been regarded as absolutely essential to the improvement of human looks, they were experts in a number of other more intricate crafts involved in beauty-culture. Their guild ran or had controlling interests in hot baths which enjoyed a very great popularity among wealthy pleasure-seekers. In the highly important traditional rôle of bath-attendants, the barbers and their kind had unique privileges. The nature of their craft made it possible for them to mix freely with the highest in social circles, and have intimate contact with the most eminent people. Plutocratic gentlemen had constant need of their services—considering it beneath their dignity to exert themselves even to tie their own turbans, they left it to their barbers to arrange their elaborate and cumbersome headgears for them. The barbers also specialized in giving sudatory treatments intended to reduce the paunch; and these, naturally, were in great demand among all those men-about-town who were a little too well-endowed with the gifts of the flesh. The dandies regularly and religiously submitted themselves to their corrective, though no doubt kindly, ministrations. They massaged the massive posteriors of the mighty with soothing unguents; pressed down the obtrusive stomachs of indolent nobles to more presentable proportions; rounded, as best they could, the unseemly angularities of aristocratic figures; applied rejuvenating unctions to senescent voluptuaries to restore to their rheumatic and dropsical limbs something of their lost youthful dexterity; rubbed tonics into bald scalps. All this, and more, was their day's work.

The fair sex depended on their craft for all those subtle touches which form an essential part of the female mystery. They supplied most of the expensive toilet-preparations for the inmates of royal seraglios, ladies of rank, wives of prominent patricians, concubines of rich merchants, and the much-sought-after courtezans—cosmetics for their lips and lustre-giving ointments for their eyelashes, scented shampoos for their hair and painstakingly distilled perfumes meant to sprinkle the warm sanctuaries of flesh and blood with the cool fragrance of flowers. In brief, the barbers, if they could not actually claim to be creators of things of beauty, could say at least that they tended beauty tenderly.

For the dealers in various luxury-commodities the Sublime was merely a business proposition, a matter of bread and butter, and a convenient way of making fortunes. But for its votaries it had an entirely different and more crucial significance. It was an important part of the technique of noble living. Leisure has always been the most tantalizing problem of people of independent means; at times it has become their nemesis. So it was in Gautama's time. The recluse, lost in the depths of his personal solitude, could spend days and nights at a stretch in an unbroken and satisfying trance of vacancy; or he could feel in turns exalted and terrified by the simple contemplation of hair on his navel. This enviable, child-like capacity for innocent wonder, this rapture of self-absorption spared him the agonies of boredom. Those in the lower walks of life were far too absorbed in making a living to have time to be bored. Only the rich were afflicted with the burden and boredom of leisure, and were forced to devise entertainments to fill empty hours.

Essentially, these entertainments were not very different from those with which the mystic brotherhood of leisure tries to kill time to-day. Perhaps, in some directions, their range was a little limited; in others, more varied. As always, there were the thrills of 'the chase.' The country was full of dense forests which abounded in game of every description, large and small. Hunting parties were a regular feature of the life of the rich. The adventures of big-game killing had a peculiar fascination for aristocrats, and young princes and nobles did not feel they had justified their claim to exalted birth until they had bagged an elephant, or a lion.

There were other less hazardous pastimes of the same order. Wrestling, boxing, and archery tournaments were held to enable the rich of both sexes to feast their eyes on public displays of exceptional physical strength. Apart from these recognized exhibitions there were combats 'of elephants, horses, buffaloes, bulls, goats, rams, cocks, quails,' and indeed almost any other living creatures which could be trained or goaded to tear one another to pieces. These spectacles were organized under the special patronage of the Court and aristocratic circles.

There were also gentler diversions. There was the familiar pageantry of State functions such as 'manœuvres, reviews, and receptions.' It not only gave the ruling oligarchy something to do, but served to sustain in the minds of common people the comforting illusion that their superiors, though they might be more fortunate in some respects, were not always idle, and had, in fact, many onerous duties to fulfil.

There were music-makers and ballet-dancers to relieve the after-dinner torpor of the rich with their ethereal arabesques of sound and motion, their evanescent patterns of harmony and poise. There were actors and ballad-singers to move the phlegmatic courtiers to tears or laughter. The promise of a few copper *kahapanas* was all the temptation they needed to create scenes of passionate love and hatred, to move themselves to the highest pitch of despair or ecstasy, to articulate the unutterable problems of life and death in ringing rhythms and sonorous phrases. There were acrobats, strolling mimes, and capering clowns to add a touch of frivolity.

The written word was not yet much in use in India, and the race of scribes had not yet come into its own. This deprived the more serious-minded among the wealthy devotees of Sublimity of the opportunity to pursue infinite knowledge in the narrow seclusion of their studies. But ways and means were found to make up for this deficiency. For their amusement and enlightenment, it was a time-honoured custom among the nobility to arrange frequent battles of wit between professional wranglers and masters of rhetoric, amateur philosophers and public orators. An original epigram, a quick repartee, a subtle spark of irony could make or mar reputations at such debates. He who could talk loudly and well, set forth long-winded arguments on abstruse problems of cosmogony, on the nature of Reality, and other kindred metaphysical matters which so troubled the contemporary mind, did not have to worry about his livelihood; he was easily able to market his talent at a fair price to some wealthy patron. Besides, as in all feudal times, the professional story-teller was a recognized institution and served the function of literature. A prodigy of memory and invention, he had developed the art of improvising stories of every variety, both in prose and verse, to a degree of accomplishment which has rarely been surpassed by sophisticated modern dealers in fiction. It was a common habit of well-to-do people to keep a story-teller as a permanent member of their household. Ladies and gentlemen of rank spent many an evening or afternoon listening to his entertaining tales 'of kings, of robbers, of ministers of state; tales of wars, of terrors, of battles of long ago; tales dealing with foods and drinks, clothes, beds, perfumes; tales of relationships, equipages, villages, towns, cities and countries; tales of ghosts, of buried treasures, of adventures and disasters on sea; tales about women and heroes and warriors long since dead; tales about things which are and things which are not.' He was their equivalent of a large and well-stocked library.

Sedentary pastimes like chess, dice, and 'games on boards with eight or ten rows of squares' filled in a few more gaps in the chart of leisured existence. We hear, too, of several other outdoor and indoor sports. Various kinds of games played with balls are recorded; and, for some reason or other, these seem to have been the favourite feminine sport. In the evenings merry matrons as well as modest girls would gather in pleasure-groves to play with balls, which, says J. J. Meyer, 'would be quite finely coloured, at any rate those of the upper classes.' Fortunately, the frolics of these tender but athletic maidens were not always confined to the strict privacy of their four walls. They took place in the open, in public parks, where many interested spectators could take pleasure in looking on. Perhaps the interest of these spectators centred less in the game than in the players, and the supple, graceful movements of 'banana-like thighs.' Judged from the lyric transports of lubricity to which it inspired the Epic poets, the spectacle must have been enthralling.

Yet all these sports only touched the fringe of the problem. They were all in the nature of mere preliminaries. The crucial rites of ecstasy were performed elsewhere; the climax, the consummation of the Sublime quest was reached in the bedchamber. 'The erotic,' says Licht, 'is the key to the understanding of Greek culture.' Indeed it is the key to the understanding of all cultures, from the most primitive to the most refined. It undoubtedly furnishes an important clue to the understanding of the complicated growth of ancient Indian civilization.

The fabric of Indian culture has always had a criss-cross pattern. If, as has often been observed, there runs through it the strand of an irrepressible, vibrant spirituality, this is intersected at every point by an equally live thread of sensuality. 'In the soul of the Indian,' writes J. J. Meyer, 'there dwells that twin pair: burning sensuality and stark renunciation of the world and of the flesh.' Yet viewed from an ultimate angle, the voluptuary and the ascetic are one: they pursue an identical object. The mystical thinkers of ancient India were fully aware of the dialectic of human experience; in their comprehensive outlook on life, they recognized that one way of attaining supreme wisdom was through the delirium of physical love. It is not only the secular literature of the Epic and Classical periods which vehemently insists on representing 'the enjoyment of women as the most glorious thing in heaven and on earth, the one meaning and end of living.' Such an emphasis would not be surprising. But what is peculiarly significant is that these healthy sentiments are shared by the compilers of our sacred literature, who stress the need of ministering to the 'Body Mystical' with the same insistence as the lay writers. 'He that is melted together with his beloved in the spell of delight . . .' says one audacious Law-Giver, 'has come into Brahma, into Nirvana . . .' The Tantric Texts go much further in audacity in these matters: their ecstatic philosophy sees physical union as 'the highest of the five things leading to perfection and the knowledge of Brahma,' and gives practical instruction for the performance of certain ingenious and elaborate sexual rites as means of access to the divine.

The extremes of self-indulgence and self-denial generally go together in human nature, and there is some justification for the belief that the Indian soul has risen to such impressive heights of renunciation precisely because of its 'burning sensuality.' What is more, 'stark renunciation of the world and the flesh' has often been an act of selfishness. The ascetic surrendered his right to the limited range of sensual pleasures on earth in order to entitle himself to the enjoyment of that fuller concupiscence which he believed to be the happy lot of denizens of the Land of Bliss. Thus his self-restraint was frequently a wise, though necessarily speculative measure of thrift: he saw 'the love of many and lovely women, shining before him as the goal and reward of his asceticism in the world beyond.' Some of our ascetics have quite frankly admitted this motive as being at the root of their asceticism. An eminent Rishi, for example, candidly confessed that having carnal knowledge of women 'was the one happiness' worth seeking, that 'redemption without it would be but threefold captivity,' and that 'it alone was real in the empty world.'

The profound undercurrent of sensuality which runs through old Indian culture is perhaps nowhere better revealed than in the *Mahabharata*. Very aptly, J. J. Meyer likens it to 'a great sea, to sail over which "threefold bronze" is needed, not indeed, about the breast, but in any case about another, less heroic part of the body.' With equal appropriateness, Macdonell describes it as 'a moral encyclopaedia.' It has the vastness of 'a great sea,' and like the kaleidoscopic depths of its waters, it mirrors life in an infinite variety of colours and forms. It possesses, too, the comprehensiveness of 'a moral encyclopaedia' and gives a faithful account of the bewildering diversity of human characters and types. It presents us men and women in their promiscuous actuality; it draws the moral ideality which creatures of flesh and blood are always painfully struggling to achieve, but for ever failing to realize; and if it contains contradictory statements regarding the moral behaviour of men and women, these are contradictions which can in fact be seen in the everyday manifestations of human nature. The unforgettable picture it creates is one of a world ethical only in appearance; where moral restrictions exist—but only to be broken; where people, though preserving a certain stiff formality without, allow themselves a comfortable laxity within. Indeed, it paints the world very much as it has been beneath the surface ever since man's fall from grace. In one of its episodes we hear the sun-god saying: 'All women and men are without restraint, O lovely-faced one. This is the real nature of mankind, any other is to speak untruly, as the holy tradition teaches.' It is true that in advancing this argument, the sun-god was actuated by rather personal motives; he was trying by such gentle persuasions to induce Kunti, the future wife of Pandu, to yield him her maidenhead (a stratagem, which, it must be remembered, achieved its end and had a very happy issue in due course of time). But even gods, like men, will admit truth only in such moments of urgency. On the whole, the behaviour of the players on the stage set up by the Epic not only corroborates the sweeping generalization of the sun-god, but to an extent vindicates his own act of seduction.

And the world of the Epic represents a fairly accurate picture of the moral conditions in which Gautama grew up to maturity. It was an age which had devoted a great deal of attention to experiment in the field of love, and developed erotic technique to the highest point of delicacy, as well as debauchery. It is not known at what exact date Vatsyayana lived, and, with a loving devotion worthy of the cause, compiled the work which to this day remains one of the most comprehensive and inspiring treatises on the theory and practice of eroticism. But the elaborate variations on the eternally fresh theme of amorousness, which the *Kama Sutra* describes with a rare wealth of imagination and detail, had doubtless been practised by men and women countless times before a man of genius came along to record them for the delight and edification of posterity. The initiates of *Ars Amandi* among Gautama's friends knew nothing if they did not know these intricate mysteries. To them love-making was not simply an entertaining and pleasurable pastime: it was an outlet for all kinds of contradictory human emotions, the quickest route to the land of heart's desire, a satisfying philosophy of life, and an extremely noble way of living.

For with Bhima in the *Mahabharata* they believed that 'without Kama a man has no wish for worldly profit, without Kama a man does not strive after the good, without Kama a man does not love'; they believed, too, that it is possible to cure the ills of the soul through an ultimate ministration of the senses and the flesh.

Red, says Frederick Duhn, is the colour of death. That may be so. But red, in all its beautiful varieties of shades, is also, and even more appropriately, the colour of life. Crimson is the tint of warm and healthy blood. Deep red roses and rubies, soft flesh-coloured carnations and carnelians are emblems soaked in the age-long symbolism of desire—of desire not in its passive languorous rôle, but in its active manifestation. In India, from the remotest antiquity to the present day, it has been a commonly accepted convention to make the bridal dress out of some vermilion stuff, as though in anticipation of lost virginity; brides on the morning after their wedding night are generally expected to produce an intimate article dipped in an even richer red, witness that they have ceased to be virgins, and have been consummated into full womanhood. And even when the maid has attained the status of a fruitful wife, this flamboyant colour continues to figure prominently in her toilet. For in India it has always been considered the special privilege of women in the married state to stain their hands and feet with blood-red henna. Also they religiously decorate their foreheads with the highly appropriate marks symbolic of their fortunate lot, drawn with the bright-red pigment obtained from cinnabar.

Thus, if in the age in which Gautama lived (as also for a long while after), that important personage of the ancient world, the courtezan, was in the habit of wearing scarlet garments it was not at all because there was some social stigma attached to her calling. Nor was it intended to mark her out 'as something criminal and ill-omened.' On the contrary, J. J. Meyer informs us, she showed preference for this vivid colour for the same reasons which, in mediaeval Germany, induced the charitable Sisters of the Order of Saint Amor to display themselves in vivid yellow; in other words, that she might 'be distinguishable for the greater ease and comfort of men.' Scarlet was her trademark; but the trade she plied was different from all other trades, the goods she dealt in had little in common with the gross commodities exchanged in ordinary commerce.

As the recognized dispenser of pleasure, the *Vécya* or harlot, was naturally one of the principal attractions of the city life in ancient India. 'In towns and cities,' writes J. J. Meyer, 'the harlot, often very wealthy and of a great distinction and quite often very well educated, went in her splendour along the street, taking to herself the fiery eyes and hearts of men . . . Ever since the dim days, when Dirghatamas, saint and poet of the Vedic songs, blind from birth, brought into the world the pleasures of love granted for ringing coin, the horizontal trade had been flourishing in the land of India.' But even this is rather an understatement regarding the immense popularity of the scarlet woman among the ancients. She was actually the nucleus of the cultural and artistic life of the community. As with the culture of Greece, so with Indian culture, much that is good and true and beautiful in it, is owed to the hetaera. Indeed she was the mother of arts. Into music and the dance, for instance, she poured her heart's blood; poetry and painting, sculpture and drama too were inspired by her to a greater or lesser degree.

In Gautama's days, the harlot enjoyed a social status very different from a woman of the streets in our time. She was not an outcast from society. Her class, it is true, occupied a somewhat lower rank in the social order—but she was not a pariah. In actual practice, indeed, she exercised an influence far in excess of the inferior place technically allotted to her on the social ladder. Men of the highest rank were at her feet, ready to serve her and satisfy her whims. Princes and poets, patriarchs and philosophers, merchants and mystics, all alike vied for her favours and were glad to do her bidding. Being herself well-equipped with physical graces as well as gifts of the spirit, she invariably attracted the best wits and intellects of the age into her winnowed circle. Witty in her conversation, wide-awake in her sympathies and interests, graceful in her movements, and highly polished in her manners, she realized in her person the qualities of Joachim du Bellay's ideal courtezan who was at once '*sage au parler, et folastre à la couche*.' The literature of the time is resonant with loud hymns in her praise; and in the delirious vision of the Epic poets she emerges 'as the very embodiment of perfect womanhood.'

It would be idle to pretend that every 'public woman' lived up to this exacting standard. There was, no doubt, the cheaper type of prostitute to suit lighter pockets. In the Athens of Socrates and Plato, we are told, there were brothels to cater for the needy at a variety of prices—starting from a minimum sum of approximately a penny and rising to fabulous figures. Similarly, in Benares, in Savatthi, and in Rajagaha, where Gautama held his discourses, there were institutions of the same kind which could satisfy widely different tastes and incomes. It must be said, however, that even these poorer harlots were not altogether devoid of a certain measure of refinement. There was the all-important fact that prostitution was considered an entirely legitimate vocation. This not only gave the demi-mondaine a high sense of self-respect and personal dignity, but also made her guard the reputation of her profession jealously. Though her interest resided mainly in clients' purses, she took at the same time a craftsman's disinterested delight in her craft, and conducted her 'horizontal trade' with a modicum of good taste. This tradition of professional good taste among the harlots, it must be recorded, was kept alive in India even until quite recent times. So weighty and experienced an authority as the eminent Abbe Dubois has observed that 'in public the Indian prostitutes behave better than their European sisters.'

The courtezan in Gautama's days was not only spared social ostracism, but she was actually received and honoured by the best society. The State, far from subjecting her to persecution, regarded her as a valuable asset and gave her all possible encouragement. And for very good reasons. Like monumental wonders and historic sights, a beautiful and accomplished courtezan attracted large crowds of curious visitors from far-off places to the township or city where she lived. Her 'invisible trade' brought prosperity to the people around her; and this in its turn served to swell the State treasury.

A story given in the Vinaya reveals the important part played by the harlot in the State economy. 'At that time,' we read, 'Vesali was an opulent, prosperous town, crowded with people, abundant with food. The secret of all these blessings enjoyed by Vesali was the courtezan Ambapalika.' For 'she was graceful, pleasant, gifted with the highest beauty of complexion, well versed in dancing, singing, and lute-playing, much visited by desirous people. She asked fifty (*kahapanas*) for one night. Through her presence Vesali became more and more flourishing.' A merchant of Rajagaha visiting Vesali on business was naturally impressed by what he saw. He went into the matter carefully, and found out who was responsible for the blissful abundance in which the inhabitants of Vesali seemed to be living. And it occurred to this patriotic merchant, that if Vesali could solve all its economic difficulties with the help of a bewitching courtezan and grow rich on the proceeds of her body, why should not Rajagaha adopt similar measures? It looked simple and reasonable.

On his return he immediately sought an interview with King Bimbisara of Magadha. The audience was granted. After giving glowing accounts of what he had seen in the rival capital, and explaining at some length what he thought were the reasons for its prosperity, he came to the main point of his argument. 'May it please Your Majesty,' he earnestly pleaded, 'let us also install a courtezan.' Bimbisara was both a merry monarch and a shrewd business man. He saw at once that the patriotic merchant was not joking. He was delighted with the suggestion which seemed to be both practical and profitable. Enthusiastically he gave royal consent to the proposal. 'Well, my good sir,' he said to the merchant, 'look for such a girl whom you can install as courtezan.' The merchant was greatly encouraged and lost no time in executing his excellent design. The Texts say: 'Now at that time there was at Rajagaha a girl Salavati by name, who was beautiful, graceful, pleasant, gifted with the highest beauty of complexion. That girl Salavati the Rajagaha merchant installed as courtezan . . . And before long she was well versed in dancing, singing, and lute-playing, much visited by desirous people.' She decided to fix the fee for her favours at a sum twice as high as that charged by Ambapalika, and asked 'a hundred *kahapanas* for one night.' The patriotic merchant and the merry and shrewd monarch had not miscalculated. Their scheme proved prodigiously successful; they both made huge profits. And not only they, but many others besides. Through the untiring exertions of this charming and gifted lady, we learn, Rajagaha also became 'an opulent, prosperous town, crowded with people, abundant with food' and even more flourishing than the capital of the rival confederation.

Such was the power of the harlot. She was more than a mere woman, she was a force of nature—irresistible, and beyond good and evil. In vain did the disgruntled priests, finding their temples half-empty at evening prayers, fulminate against her in their bitterness. In vain did they invoke the wrath of gods and turn upon her the battery of their 'characteristic abuse, copious and not at all stinted.' And because they were angry, in their invective they were often at a loss and mixed their metaphors. 'As bad as ten slaughter-horses,' they raved in their impotent rage, 'is an oil-miller's wheel, as bad as ten oil-millers' wheels is an inn-sign, as bad as ten inn-signs is a harlot.' Unapprehending but intimidated, their audience tried in vain to puzzle out the similitude between a harlot and a slaughter-horse, and seeing no nexus between the two, shook their heads in consternation. In vain did the moralists, realizing that they were doomed to have a lonely journey along the straight and narrow path, inveigh against the depravity of the age, raise their fists menacingly against the laity who were constantly slipping out of stiff-looking moral garments into the soft depths of the harlot's bed, and move heaven and earth in an attempt to wipe out the abomination of whores and whoredom. Their vituperations, their reproaches, and even their threats availed nothing; or, worse still, recoiled with ridicule on their own heads, already bald through excessive care for the well-being of an ungodly and ungrateful age.

The vogue of venal love flourished unchecked. Beyond good and evil, beyond the reach of the splenetic breath of priestly and prophetic censure, the scarlet woman remained unmoved on her throne of glory, presiding like a benevolent deity over endless festivals of fire. When she saw the upholders of morality trying to reduce her to ashes with their fierce looks, she, who was herself of the very nature of the flame, smiled a meaningful smile—a smile of understanding and pity, of forgiveness and promise. And they who reviled her so ferociously had need of all these great mercies. For every now and then, under the inexorable pressure of human necessity, they, too, found themselves knocking at her door, if not exactly in sackcloth, at least with their remorseful eyes averted and downcast—the eternal prerogative of the penitent. And she took in these prodigal children, offered them the warm comfort of her body, and relieved them of the painful burden of desire—and the weight of their wallets. The scarlet woman could afford to be generous—even to those who were her avowed enemies. Perhaps, through some inscrutable intuition, she could sense that they who pretended so heartily to abominate her were at heart themselves in love. She herself was free from the tentacles of love and hate—beyond good and evil.

It was in such an atmosphere of sensuality and refinement, indolence and intellectual liveliness, high culture and lavish luxury that the early years of Gautama's life were spent. With a curious though understandable vehemence, the chroniclers insist that during this period of his career the Tathagata indulged in a more than moderate measure of 'dalliance in love.' Indeed, the legend is even more explicit regarding his early amorous adventures than his later, more serious metaphysical activities. Richly coloured accounts try imaginatively to recreate every possible detail of Gautama's youthful escapades. A page or two of Asvaghosha's poem illustrates the keen delight taken by the Classical biographers in conjuring these amatory scenes:

'Then surrounded by troops of women the prince wandered in the wood like an elephant in the forests of Himavat accompanied by a herd of females.

'Attended by the women he shone in that pleasant grove, as the sun surrounded by Apsaras in his royal garden.

'There some of them, urged by passion, pressed him with their full firm bosoms in gentle collisions.

'Another violently embraced him after making a pretended stumble—leaning on him with her shoulders drooping, and with her gentle creeper-like arms dependent.

'Another with her mouth smelling of spirituous liquor, her lower lip red like copper, whispered in his ear, "Let my secret be heard."

'Another, all wet with unguents, as if giving him her command, clasped his hand eagerly and said, "Perform thy rites of adoration here."

'Another with her blue garments continually slipping down in pretended intoxication, stood conspicuous with her tongue visible like the night with its lightning flashing.

'Others, with their golden zones tinkling, wandered about here and there showing to him their hips with veiled cloth.

'Others leaned, holding a mango-bough in full flower, displaying their bosoms like golden jars.

'Another, coming from a lotus-bed, carrying lotuses stood like the lotus-goddess Padma, by the side of that lotus-faced prince.

'Another sang a sweet song easily understood, and with the proper gesticulations, rousing him, self-subdued though he was, by her glances, as saying, "Oh how thou art deluded!"

'Another, having armed herself with her bright face, with its brow-bow drawn to its full, imitated his action, as playing the hero.

'Another, with beautiful full bosoms, and having her earrings waving in the wind, laughed loudly at him, as if saying, "Catch me, sir, if you can!"

'Some, as he was going away, bound him with strings of garlands—others punished him with words like an elephant-driver's hooks, gentle yet reproachful.

'Another, wishing to argue with him, seizing a mango-spray, asked, all bewildered with passion, "This flower, whose is it?"

'Another, assuming a gait and attitude like those of a man, said to him, "Thou who art conquered by women, go and conquer this earth."

'Then another with rolling eyes, smelling a blue lotus, thus addressed the prince with words slightly indistinct in her excitement: "See, my lord, this mango covered with its honey-scented flowers, where the kokila sings, as if imprisoned in a golden cage.

' "Come and see this asoka tree, which augments lovers' sorrows—where the bees make a noise as if they were scorched by fire.

' "Come and see this tilaka tree, embraced by a slender mango-branch, like a man in a white garment by a woman decked with yellow unguents.

' "Behold this kuruvaka flower, bright like a lip which has put pigments to shame.

' "Come and see this asoka tree, covered all over with new shoots, which stands as it were ashamed at the beauty of our hands.

' "See this lake surrounded by the sinduvara-shrubs growing on its banks, like a fair woman reclining, clad in fine white cloth.

' "See the imperial power of females—yonder ruddy-goose in the water goes behind his mate following her like a slave.

' "Come and listen to the notes of this intoxicated cuckoo as he sings, while another sings as if consenting, wholly without care.

' "Would that thine was the intoxication of the birds which the spring produces—and not the thought of a thinking man, ever pondering how wise he is!" '

On another occasion, Asvaghosha shows the Bodhisattva being 'perforce carried away to a wood filled with troops of beautiful women, just as if some devotee who had newly taken his vow were carried off, feeling weak to withstand temptation, to the palace of the monarch of Alaka, gay with the dancing of the loveliest heavenly nymphs.' These, and a multitude of other similar descriptions, are, of course, the product of the novelist's touch. But allowing for Asvaghosha's 'writing up' of the story, and ignoring his tiresome far-fetched similes and metaphors, there still remains a kernel of fact to his fiction. His biography merely represents Gautama as living the kind of life which was actually being lived by the upper strata of Indian society of his time. Being the spiritual adviser of king Kanishka, Asvaghosha had ample opportunity to observe the habits and manners of the privileged classes. He had frequently seen the scions of wealthy families disporting themselves with 'troops of beautiful women' in the woods. He had known princes who were in the habit of finding relaxation from their exhausting princely duties in the arms of full-bosomed concubines. He was not unaware that the patriarchal harems were overcrowded. He had witnessed his own king being assailed by the passionate women of his seraglio 'with all kinds of stratagems.' The scene in which he moved in the first century A.D. he simply transferred to the sixth century B.C. But this does not mean that the picture he paints is altogether devoid of reality. On the contrary, in all its incidental details it remains essentially a true record. The *mise en scène* in Asvaghosha's India did not differ materially from the stage on which Gautama had played his part six centuries earlier.

It is doubtful if Gautama himself ever felt really quite at home in the milieu in which he was born. The orgies of sensual indulgence, which were evidently a matter of daily routine among the members of his privileged class, were hardly suitable to his contemplative temperament. By nature and inclination, he was more fitted to be a spectator than an actor in these delightful frolics. His self-consciousness—which in his adolescence earned for him the reputation of being 'lazy'—probably also deprived him of the enviable capacity for that spontaneous 'intoxication of the birds which spring produces'; and he was perhaps both too self-possessed and undemonstrative to play the game of hearts with any great measure of success.

This much may be gathered from hints dropped by the writers of various accounts of his life. Asvaghosha emphasizes that the Bodhisattva was inclined to be somewhat 'self-subdued' and 'sedate.' Further he admits that, even before the great crisis of his life came, 'the prince' found no particular pleasure 'in the women's apartments, in the several objects of the senses,' and the spectacle of redolent and resilient female flesh, with its tempting promise of 'pneumatic bliss,' merely made him think of the fickleness of youth and beauty, of the transiency of things. Even 'the sweet sounds and the rest,' he laments, served to evoke in Gautama's heart a thought of grief—the grief born of a realization of the inevitability of old age, death, and decay.

It is true that in emphasizing all these attributes Asvaghosha was trying to create a character that would closely conform to the conventional type cast for Bodhisattvas—'those beings of pre-eminent nature who, after knowing the flavour of worldly enjoyments,' are in the habit of departing 'to the forest as soon as a son is born to them . . . to attain supreme wisdom.' But, at the same time, it is conceivable that this particular convention set up by tradition was originally based on idealization of certain conspicuous traits actually observed in Gautama's personality by his contemporaries. It can thus be regarded as possessing some degree of verisimilitude. The love-sick maiden in Asvaghosha's account who, gently, 'in words slightly indistinct with excitement,' takes Gautama to task for having 'the thought of a thinking man, ever pondering how wise he is,' shows a very real understanding of his character. Judged from all accounts, Gautama appears to have been the kind of a person who is fated to suffer a life-long crucifixion in himself, and for whom the condition of personal isolation, though no doubt a painful burden, represents the only possible means of personal integration. Even from the early years of his youth, the very intensity of his sensitiveness had created an impenetrable barrier around him; and it is quite likely that he found it difficult to yield 'wholly without care' to the lure of lovely female forms in their tempting abandonment and surrender, or to participate wholeheartedly in the pleasant rites of virgin adoration. For such an individual even the possibility of these human contacts is greatly reduced. Pierced by the sharp arrows of his own inscrutable thoughts and feelings, possessed of depths which were beyond the reach of his fellow beings, and haunted by yearnings unknown and incomprehensible to his most intimate companions, Gautama probably had no choice but to be—though not in the romantic sense which Rousseau certainly meant to give the phrase—a '*promeneur solitaire.*'

However, the most self-conscious spectator is apt at times to forget his detached rôle and become an actor in the play which he is watching; and even those who are in the habit of walking carefully on the edge of life are occasionally drawn into its magnetic vortex. Both the spirit and the flesh are weak. And the temptation not to miss opportunities of any kind is very great. For a man placed as Gautama was, the temptation was still stronger. The canonical poets singing of the beauty and enchantment of the mythical city of Kusavati where the Great King reigned in all his glory say: 'Both by day and night . . . the royal city of Kusavati resounded with ten cries; that is to say, the noise of elephants, and the noise of horses, and the noise of chariots; the sounds of the drum, of the tabor, and of the lute; the sound of singing, and the sounds of the cymbal and of the gong; and lastly, with the cry, "Eat, drink, and be merry!" ' Kapilavastu was not, of course, the royal city of Kusavati; but it was not altogether a stranger to these inebriating and festive cries. And in such a joyous atmosphere even a Bodhisattva might, for a few brief moments at any rate, be tempted to enjoy the fullness of earthly bliss and forget the cankerous grief gnawing at his soul.

V

THE CRISIS AND RENUNCIATION

I have of late, (but, wherefore, I know not,) lost all my mirth . . .

Hamlet

BUT there is a worm in the bud which slowly eats the heart out of those certitudes upon which the world so fondly builds its hopes and loves. It is true that nature, by way of a defensive mechanism, equips us with a veil of ignorance which effectively conceals from our eyes the depths of uncertainty on the brink of which, like so many somnambulists, we are always walking. It is true that the stuff of which this veil is made possesses quite extraordinary thickness and resiliency. Nevertheless, the veil of our ignorance is not altogether shock-proof: it has been known to give way under the impact of reality. And at the root of the human mind there is also the restless germ of doubt; doubt which cuts deeper than make-believe; doubt which refuses to take things for granted; doubt which is impatient of camouflage; doubt which is at once a liberating and a tragic element. Liberating, because it is only through the negation of what is commonly taken for granted, of what is conventional, that human awareness can hope to expand itself—can hope to integrate itself in ever-widening patterns of perception. Tragic, because the release of consciousness itself imposes a stress on the experiencing mind; because it opens up to human vision unknown horizons where man is apt to lose his bearings. Liberating, because doubt is a vehicle of light. Tragic, because light is a destructive force—destructive of the myths and illusions which constitute the psychological crutches of mankind. And to abandon these psychological supports is to risk a breakdown. And often there is a breakdown. Such is the tragic paradox.

This breakdown is not necessarily a spectacular affair; it is merely the recognition of a failure—the failure of an individual mind to bear the burden of its awareness. Tragedy is not a melodrama; it is the realization of a particular relationship—specifically, a negative relationship—between the subjective world of experience and the objective conditions of existence. The tragic experience need involve no blood baths and sanguinary convulsions: at its most tragic, it is simply the apprehension of a condition of deficit and want so immediate as to be almost unutterable. In more than one sense can be attributed to the tragic experience all those paradoxical qualities which the Buddhists attribute to their nirvanic state: it, too, is 'profound, difficult to realize, hard to understand . . . not to be grasped by mere logic, subtle, comprehensible only to the wise.' Such essentially seems to have been the crisis that so transformed the course of Gautama's life. It was not a dramatic crisis, but a traumatic crisis of the human awareness itself. Consequently, it would be futile to look for a connected series of events dramatically culminating in Gautama's decision to renounce his worldly estate, his home, his parents, his wife and child, and to go forth to the life of a homeless wanderer. The explanation of the causes which led up to this climax, if there is one to be found at all, has to be sought in some kind of a deeper crisis of the mind.

As usual, the legend dramatizes. However, two subjective motifs distinctly emerge from the dramatic narrative; both bear a curious and striking resemblance to the motifs which recur with pathetic emphasis in *Ecclesiastes.* There is the motif of world-weariness arising from a realization of the 'wearisome conditions of humanity, born under one law, to another bound.' Like the Preacher, Gautama is shown as being terribly obsessed with thoughts of the misery inherent in human life—the misery of birth and death, of old age and decay. The Preacher had lamented:

'I made me great works; I builded me houses; I planted me vineyards:

'I made me gardens and orchards, and I planted trees in them of all *kind of* fruits:

'I made me pools of water, to water therewith the wood that bringeth forth trees:

'I got *me* servants and maidens, and had servants born in my house; also I had great possessions of great and small cattle above all that were in Jerusalem before me:

'I gathered me also silver and gold, and the peculiar treasure of kings and of the provinces: I gat me men singers and women singers, and the delights of the sons of men, *as* musical instruments, and that of all sorts.

'So I was great, and increased more than all that were before me in Jerusalem: also my wisdom remained with me.

'And whatsoever mine eyes desired I kept not from them, I withheld not my heart from any joy; for my heart rejoiced in all my labour: and this was my portion of all my labour.

'Then I looked on all the works that my hands had wrought, and on the labour that I had laboured to do: and, behold, all *was* vanity and vexation of spirit, and *there was* no profit under the sun.

'And I turned myself to behold wisdom, and madness, and folly: for what *can* the man *do* that cometh after the king? *even* that which hath been already done.

'Then I saw that wisdom excelleth folly, as far as light excelleth darkness.

'The wise man's eyes *are* in his head; but the fool walketh in darkness: and I myself perceived also that one event happeneth to them all.

'Then said I in my heart, As it happeneth to the fool, so it happeneth even to me; and why was I then more wise? Then I said in my heart that this also *is* vanity.

'For *there is* no remembrance of the wise more than of the fool for ever; seeing that which now *is* in the days to come shall all be forgotten. And how dieth the wise *man?* As the fool.

'Therefore I hated life; because the work that is wrought under the sun *is* grievous unto me: for all *is* vanity and vexation of spirit.'

In words slightly less rueful but charged with the same sense of affliction and self-pity characteristic of a prolonged adolescence, the Tathagata is made to declare in one of the dialogues:

'I was delicate, O monks, extremely delicate, excessively delicate. In my father's dwelling lotus-pools had been made, in one blue lotuses, in another red, in another white, all for my sake. I used no sandal-wood that was not of Benares, my dress was of Benares cloth, my tunic, my underrobe, and cloak. Night and day a white parasol was held over me so that I should not be touched by cold or heat, by dust or weed or dew.

'I had three palaces, one for the cold season, one for the hot, and one for the season of rains. Through the four rainy months, in the palace for the rainy season, entertained by female minstrels I did not come down from the palace; and as in the dwellings of others food from the husks of rice is given to the slaves and workmen together with sour gruel, so in my father's dwelling rice and meat was given to the slaves and workmen.

'Then, O monks, did I, endowed with such majesty and such excessive delicacy, think thus: "An ignorant, ordinary person, who is himself subject to old age, not beyond the sphere of old age, on seeing an old man is troubled, ashamed, and disgusted, extending the thought to himself. I too am subject to old age, not beyond the sphere of old age, and should I, who am subject to old age, not beyond the sphere of old age, on seeing an old man be troubled, ashamed and disgusted?" This seemed to me not fitting. As I thus reflected on it, all the elation in youth utterly disappeared.'

The dialogue adds that similar reflections on death and sickness still further alienated the Tathagata's mind from life. On the basis of this, the commentators and chroniclers have built up the dramatic story of the four signs. It would be quite sufficient to quote Asvaghosha's description of Gautama's encounter with a corpse:

'But as the king's son was thus going on his way, the very same deities [that is, the deities who had, on two previous occasions, arranged for an old man and a sick man to meet Gautama in order to awaken him to the fact of suffering that is implicit in life] created a dead man, and only the charioteer and the prince, and none else, beheld him as he was carried along the road.

'Then spoke the prince to the charioteer, "Who is this borne by four men, followed by mournful companions, who is bewailed, adorned but no longer breathing?"

'Then the driver—having his mind overpowered by the gods who possess pure minds and pure dwellings—himself knowing the truth, uttered to his lord this truth also which was not to be told:

' "This is some poor man who, bereft of his intellect, senses, vital airs and qualities, lying asleep and unconscious, like mere wood or straw, is abandoned alike by friends and enemies after they have carefully swathed and guarded him."

'Having heard these words of the charioteer he was somewhat startled and said to him, "Is this an accident peculiar to him alone, or is such the end of all living creatures?"

'Then the charioteer replied to him, "This is the final end of all living creatures; be it a mean man, a man of middle state, or a noble, destruction is fixed to all in this world."

'Then the king's son, sedate though he was, as soon as he heard of death, immediately sank down overwhelmed, and pressing the end of the chariot-pole with his shoulder spoke with a loud voice:

' "Is this end appointed to all creatures, and yet the world throws off all fear and is infatuated! Hard indeed, I think must the hearts of men be, who can be self-composed in such a road.

' "Therefore, O charioteer, turn back our chariot, this is no time or place for a pleasure-excursion; how can a rational being, who knows what destruction is, stay heedless here, in the hour of calamity?" '

Various other early biographers of Gautama tell us in more or less identical language how he arrived at the melancholy conclusion which Schopenhauer echoes in his *Counsels and Maxims* when he says: 'It is clear that our walking is admittedly nothing but a constantly-prevented falling, so the life of our bodies is nothing but a constantly-prevented dying, an ever-postponed death.'

The second motif on which the legend bases itself is an overtone of recoil from the physical side of life, and in particular from sex. It has already been pointed out that practically the whole of the Buddhist literature manifests a tortured preoccupation with things of the flesh. This obsession expresses itself either as an almost frenzied disgust of the body, its functions and wants, or, antithetically, as a nostalgic yearning for carnal things. Thus the chief crisis of Gautama's life is represented as a crisis of sexual disgust. The Preacher had been content merely to say: 'And I find more bitter than death the woman, whose heart *is* snares and nets, *and* her hands *as* bands: whoso pleaseth God shall escape from her; but the sinner shall be taken by her.' But the hero of Asvaghosha's sentimental poetical novel out-preaches the Preacher. There is a description of an argument between Gautama and Udayin, 'the wise son of the family priest' and a man 'well skilled in the rules of policy.' The latter tries to urge Gautama to drown his grief in that exquisite fountain which Solomon in his Song of Songs had likened to 'a well of living waters . . . a cup *which* wanteth not liquor.' 'It is right,' he argues with great adroitness, 'to woo a woman even by guile—this is useful both for getting rid of shame and for one's own enjoyment.' Of such, he adds, the lives of all great gods and heroes alike remind us. For:

'Knowing that pleasure was the best of objects, even the god Indra wooed in olden times Ahalya the wife of saint Gautama.

'So too Agastya wooed Rohini, the wife of Soma; and therefore, as Sruti saith, a like thing befell Lopamudra.

'The great ascetic Vrishaspati begot Bharadvaga on Mamata the daughter of the Maruts, the wife of Autathya.

'The Moon, the best of offerers, begat Buddha of divine nature on the spouse of Vrihaspati as she was offering a libation.

'So too in old time Parasara, overpowered by passion on the bank of the Yamuna, lay with the maiden Kali who was the daughter of the son of Water [Agni].

'The sage Vasistha through lust begot a son Kapingalada on Akshamala a despised low-caste woman.

'And the seer-king Yayati, even when the vigour of his prime was gone, sported in the Kaitraratha forest with the Apsaras Visvaki.

'And the Kaurava king Pandu, though he knew that intercourse with his wife would end in death, yet overcome by the beauty and good qualities of Madri yielded to the pleasures of love.

'And so Karalaganaka, when he carried off the Brahman's daughter, incurred loss of caste thereby, but he would not give up his life.'

In answer to these well-meant persuasions, Gautama confesses his familiar apprehensions about old age, disease, and death. 'But I am fearful and exceedingly bewildered,' he says, 'as I ponder the terrors of old age, death, and disease; I can find no peace, no self-command, much less can I find pleasure, while I see the world as it were ablaze with fire.' However, he also hints at another more subtle reason for his reluctance to bathe in the 'well of living waters.' 'Are not men unfit for women to look at and women for men?' he asks. This motif becomes still more obvious in another scene, where Asvaghosha describes the women attendants of 'the prince' lying in a state of drowsy abandonment after a late night revel:

'Another, with her hair loose and dishevelled, and her skirts and ornaments fallen from her loins, lay with her necklace in confusion, like a woman crushed by an elephant and then dropped.

'Others, helpless and lost to shame, though naturally self-possessed and endowed with all graces of person, breathed violently as they lay and yawned with their arms distorted and tossed about.

'Others, with their ornaments and garlands thrown off—unconscious, with their garments spread out unfastened—their bright eyes wide open and motionless—lay without any beauty as if they were dead.

'Another, with fully-developed limbs, her mouth wide open, her saliva dropping and her person exposed, lay as though sprawling in intoxication—she spoke not, but bore every limb distorted.

'Thus that company of women, lying in different attitudes, according to their disposition and family, bore the aspect of a lake whose lotuses were bent down and broken by the wind.

'Then having seen these young women thus lying distorted and with uncontrolled gestures—however excellent their forms and graceful their appearance—the king's son felt moved with scorn.

' "Such is the nature of women, impure and monstrous in the world of living beings; but deceived by dress and ornaments a man becomes infatuated by a woman's attractions.

' "If a man would but consider the natural state of women and this change produced in them by sleep, assuredly he would not cherish his folly; but smitten from right will he succumbs to passion." '

Evidently, the Buddhist poet and historian shared the sentiments of the German philosopher who, nearly two thousand years later, declared: 'It is only a man whose intellect is clouded by his sexual impulse that could give the name of the *fair sex* to that undersized, narrow-shouldered, broad-hipped, and short-legged race'; and bitterly inveighed against the inequity of Nature which tricks man so well that he falls inextricably into the snares set for him by the 'unaesthetic sex.'

Gautama's sexual disgust is even more strongly emphasized in a passage of the Jataka which describes the same scene as above in slightly different words, and has a peculiarly Baudelairean flavour: it evokes something of the horrible fascination of a poem like *Les Métamorphoses du Vampire:*

'Thereupon women clad in beautiful array, skilful in the dance and song, and lovely as Deva-Maidens, brought their musical instruments, and ranging themselves in order, danced, and sang, and played delightfully. But the Bodhisattva, his heart being estranged from sin, took no pleasure in the spectacle, and fell asleep. And the women saying, "He for whose sake we were playing has gone to sleep. Why should we weary ourselves?" laid aside the instruments they held and lay down to sleep. Lamps fed with sweet-smelling oil were burning. The Bodhisattva waking up, sat cross-legged on the couch, and saw those women with their musical instruments laid aside and sleeping—some drivelling at the mouths, spittle-besprinkled, some grinding their teeth, some snoring, some muttering in their sleep, some gaping, and some with their dress in disorder—plainly revealed as horrible sources of mental distress.

'Seeing this change in their appearance, he became more and more dissatisfied with sense-desires. To him that magnificent apartment, as splendid as Sakka's residence, began to seem like a charnel field full of corpses, like a great area laden with diverse offal. Life, whether in the worlds subject to passion, or in the formless worlds, seemed to him like staying in a house that had become the prey of devouring flames. An utterance of intense feeling broke from him: "It all oppresses me! It is intolerable!" '

This repugnance of a misogynist for sex and the objects of sexual desire rises at times to an unbearable pitch and achieves a positively psychopathic intensity. The disgust then ceases to be directed against women as such: it transforms itself into an hallucinated abhorrence of the human body and everything connected with it. Thus in the *Gradual Sayings* there is a chapter of the *Book of the Nines* entitled "A Boil," which, for sheer morbidity, compares favourably with the inspired utterances of some of the early Christian saints and martyrs:

'Imagine monks, a boil, which has been gathering for

many years. It might have nine gaping wounds. Thence whatever might ooze out, foulness would ooze out, stench would certainly ooze out, loathesomeness would certainly ooze out; whatever might be discharged, foulness would certainly be discharged, stench would certainly be discharged, loathesomeness would certainly be discharged.

' "A boil"—that is the name for the body, monks, made up as it is of four elements, begotten of mother and father, a lump of gruel and sour milk, impermanent, subject to erasion, abrasion, disruption, and dissolution, with nine gaping wounds, nine natural openings. And from it whatever might ooze out, foulness would ooze out, stench would certainly ooze out, loathesomeness would certainly ooze out; whatever might be discharged, foulness would certainly be discharged, stench would certainly be discharged, loathesomeness would certainly be discharged. Wherefore monks, be ye disgusted with the body.'

But the world of the legend is strictly subject to the law of opposites. Every element in it inevitably produces its antithesis. The element of recoil brings into play a powerful element of attraction; the touch of disgust is counterbalanced by an irresistible craving and desire. On the one hand, Gautama's biographers credit him with a supreme loathing for the body; on the other, they attribute to him a rueful nostalgia for its warm, sensuous delights. In the story of the Great King of Glory they show him dilating on the voluptuous beauty of his 'Woman-Treasure' in a passionate retrospect:

'Now further, Ananda, there appeared to the Great King of Glory the Woman-Treasure, graceful in figure, beautiful in appearance, charming in manner, and of the most fine complexion; neither tall, nor very short; neither very stout, nor very slim; neither very dark, nor very fair; surpassing human beauty, she had attained unto the beauty of the gods.

WOOD NYMPH
(Sanchi—*Archives of the Museé Guimet*)

'The touch too, Ananda, of the skin of that wondrous Woman was as the touch of cotton or of cotton wool: in the cold her limbs were warm, in the heat her limbs were cool; while from her body was wafted the perfume of sandal wood and from her mouth the perfume of lotus.

'That Pearl among Women too, Ananda, used to rise up before the Great King of Glory, and after him retire to rest; pleasant was she in speech, and ever on the watch to hear what she might do in order so to act as to give him pleasure.

'That Pearl among Women too, Ananda, was never, even in thought, unfaithful to the Great King of Glory—how much less then could she be so with the body!

'Such, Ananda, was the Pearl among Women who appeared to the Great King of Glory.'

Again in a Siamese *Life of Buddha* translated by Mr Alabaster in his book on Siamese Buddhism (*The Wheel of Law*), we read that when Gautama founded his kingdom of Righteousness: 'The evening was like a lovely maiden; the stars were the pearls upon her neck; the dark clouds her braided hair; the deepening space her flowing robe. As a crown she had the heavens where the angels dwell; these three worlds were her body; her eyes were the white lotus flowers which open to the rising moon; and her voice as it were the humming of the bees. To do homage to the Buddha and to hear the first preaching of his word, this lovely maiden came.'

And there are moments when this strangely satisfying nostalgia for female flesh breaks out into a song of exquisite beauty and delicacy, with all the tenderness of a love-sick heart, and poignant with the wistfulness of a frustrated desire:

'Monks, I know of no other single form by which a man's heart is so enslaved as it is by that of a woman. Monks, a woman's form obsesses a man's heart. Monks, I know of no other single sound by which a man's heart is so enslaved as it is by the voice of a woman. Monks, a woman's voice obsesses a man's heart. Monks, I know of no other scent . . . savour, touch, by which a man's heart is so enslaved as it is by the scent . . . savour, and touch of a woman. Monks, a woman's scent, savour, and touch obsess a man's heart.'

This interpretation is illuminating. It is illuminating not so much for the light it throws on Gautama's own crisis, but because it reveals unconsciously the whole psychology of the Buddhist monks. It tells us little directly of Gautama's personal experience, but a great deal of the experience of his biographers. For what it represents as Gautama's, is merely a reflection of his biographers' frame of mind. They have projected their own fears and obsessions, their own reactions and cravings on to Gautama. The motives by means of which they resolve the crisis of Gautama's life are actually those which played a most crucial part in determining their own peculiar, but not incomprehensible, psychology. Constant terror of death and decay, alternating convulsions of what Baudelaire called '*le dégoût irrésistible*' and an equally unappeasable desire for the tenderness of the things of the flesh, were very real and tremendous factors in their experience. And through an easy process of imaginative identification they concluded that such must also have been the experience of Gautama. The fallacy is obvious.

And yet, in spite of its misrepresentations and fallacies, their interpretation offers a basis for understanding the nature of Gautama's crisis—indeed, there is no other basis. This may sound paradoxical; but unless the paradox is understood and accepted there is no possibility of our gaining any insight into Gautama's life. For there is no other means of access to his personality except the legend; to reject the legend altogether would involve an undertaking to create a character out of a void. And, actually, there does not seem to be any justification for totally rejecting the evidence of the legend. It is true that its evidence is mixed with much that is falsified and untrue. Yet, often the falsity it contains is not so much factual as a subjective distortion of facts. It is very essential to grasp the difference between these two distinct forms of untruth.

Thus the accounts of the Buddhist hagiographers are probably quite correct in so far as they insist on Gautama's curious preoccupation with the problem of death, old age, and decay. It is possible that in autobiographical moods Gautama admitted to companions that before he renounced his worldly state the problem of death and decay had been an obsession with him for many years. It is possible also that he vehemently stressed these cardinal facts of human life in order to impress upon his audience their reality. But neither his immediate disciples, nor the scribes who nearly two hundred years after his death began recording the various traditions current about his life and work, could understand the real meaning of his obsession. If they could understand it at all, it was by interpreting it in terms of their own crude apprehensions. They could perfectly understand and sympathize with a person living in mortal fear of the spectre of death, haunted as they themselves were by its terrifying shadow. But a subtler recognition of the problem was entirely beyond the compass of their minds.

It was, however, not beyond the compass of Gautama's mind. Indeed, there are reasons to believe that this subtler recognition did enter into Gautama's preoccupation with the problem of death and decay. An obvious inconsistency in the legend itself renders it impossible to explain Gautama's crisis as one of fear. On the one hand, we are shown a man who finds the spectacle of a corpse so overwhelming that he faints and bemoans: 'I am fearful and exceedingly bewildered, as I ponder the terrors of old age, death, and disease; I can find no peace, no self-command, much less can I find pleasure, while I see the world as it were ablaze with fire.' On the other hand, we are asked to believe that this man formulated a view of life which offered little compensation or consolation for the insufficiency of human experience in a blissful hereafter. These two pictures can hardly be reconciled. We have to reject one or the other; and on the whole there is more justification for rejecting the former than the latter. For the more authentic and intelligent Buddhist sources make it quite clear that the desire for 'extension in time'—as Krishnamurti so aptly describes the human craving after personal immortality—which mankind finds so painful and difficult to renounce, was actually renounced by Gautama. This is highly significant and revealing. At least, it safely disposes of the 'fear' theory. For it is psychologically impossible for one so terrified of death to accomplish a renunciation of such an ultimate and difficult nature, and to give up the tempting hope of eternal life.

But the question still remains. Why was Gautama so profoundly preoccupied with the problem of death, and what were his reasons for his constant insistence that 'decay is inherent in all component things'? The answer, perhaps, is that this obsession and emphasis were a part of the method which he considered absolutely essential to a true valuation of life. Obviously, there can be no realistic valuation of life until the fact of death has been grasped in the right perspective. We cannot claim honestly to have faced the issue until we have recognized the actuality of death in its utter nakedness. In fact, the recognition of the problem truly dawns on us only when we have rid our minds of the all-too-common habit of treating life as something everlasting, regarding death as a return of the prodigal to the bosom of his heavenly father, 'who is our home.' It is easy enough to live happily, to look upon one's world as the best of all possible worlds, to take things for granted, and to face death heroically—or, at least, with a semblance of equanimity—when at the back of one's mind there is the soothing assurance that death is not really the end of one's being, but the portal to a new, and possibly better, richer life. 'Obstinate questionings of sense and outward things,' painful misgivings and despair appear only when we have discarded all 'intimations of immortality,' abandoned infantile hopes, and clearly realized that death signifies an absolute dissolution of the very thing we cherish most—our personal, individualized self—and a final setting of 'the soul that rises with us, our life's star.'

There we must seek the clue to an understanding of the crucial experience that led to Gautama's great renunciation. His crisis was not just a peculiar case of melancholia arising from the strong reaction of a somewhat hypersensitive youth to certain unpleasant facts of human life: it was something of much more vital significance. We see in his experience the birth-pangs of a new, more realistic form of awareness; a highly enlightened individual struggling to demolish for himself the hopeless, yet somehow strangely universal, illusion of the permanence of this world of sense and succession—and above all, the illusion of the permanence of the human ego. It is never an easy illusion to demolish, because as Schopenhauer wisely observes, 'we think unwillingly of things which powerfully injure our interests, wound our pride, or interfere with our wishes'; and, manifestly, there is nothing which injures our interests, wounds our pride, and interferes with our wishes quite so fatally as the menace of death and decay. So much Gautama had come to realize in the critical period of his youth.

Step by step he had been moving towards the conviction that the processes of dissolution and evolution, of becoming and breaking-up, are themselves inherent in life; that the world we live in is a world of evanescence and transiency, where everything that ever sees the light of day is bound by its own nature eventually to cease being. And in the end, perhaps, this conviction had come to him like a sudden pang. After that moment, it was natural that he should have found it difficult to take the common valuation of life for granted, or to silence with conventional evasions the questioning voice within him. He had been a householder, a husband and a father; and he did not feel that he had been very successful in fulfilling the exacting demands of these highly respectable patriarchal rôles. He had seen the life of sensual indulgence and abandonment, of luxury and moral laxity; and he had discovered its utter futility. He had heard the tempting cry of 'eat, drink, and be merry'; and he had found that at heart it was a hollow cry, signifying nothing—nothing at all. This was one aspect of his crisis.

There was also another side to it. The insistence of the Buddhist texts on Gautama's sexual recoil has to be taken into account. Here again, it appears, we have the vulgarization of a profound fact of human experience. There is more in it than the mere recoil of an adolescent consciousness from the physical manifestations of desire. It is quite possible that Gautama did say: 'Monks, I know of no other single form by which a man's heart is so enslaved as it is by that of a woman. Monks, a woman's form obsesses a man's heart,' or words to that effect. But there is nothing in this discourse to suggest that convulsive disgust and loathing of an inverted voluptuary for the flesh, and in particular for the flesh of a woman, which seems to inspire most of Gautama's canonical as well as post-canonical biographers. Gautama's obsession, in so far as it was an obsession at all, appears to have been born of an awareness of a fundamental psychological fact—the tyranny that is implicit in all desire and passion. Very early in his life he had come to realize that there is no greater and more inexorable bondage than the bondage of one's own insatiable cravings. He had seen the misery and wretchedness of what human beings describe as love, friendship, and tenderness; seen, too, that these seemingly exalted sentiments were in reality no more than thinly disguised forms of sentimentality, selfishness and concupiscence. And, rightly or wrongly, he felt that sentimentality, selfishness, and concupiscence did not afford sufficient sanction for human relationships; that to be worth while, these relationships had to be divested of their sordidness and lifted to an altogether more noble, innocent, and impersonal basis. His recoil, if it was recoil, was symptomatic of a struggle within his own being to achieve a new innocence and purity.

The crisis which led to his renunciation, and urged him to set out on his strange quest, can be understood only in terms of these intricate psychological factors. At the time he left his home Gautama might not have been fully aware of the compulsions at work in his soul. The whole problem probably defined itself to him in the process of the search. But there is no reason to suppose that he took the drastic step of renunciation altogether in the dark. He must have known, even if somewhat hazily, the main object of his quest; known that he wanted to discover whether

> *Derrière les ennuis et les vastes chagrins*
> *Qui chargent de leur poids l'existence brumeuse*

there was anything in the nature of a hope that was not a day-dream, a fulfilment that was not a conscious or semi-conscious attempt at self-delusion, a human relation that was not based on possession—and finally, a tenderness that was not tainted with sentimentality, selfishness, or concupiscence.

Of the incidents connected with Gautama's going forth from home, very little is known which can be described as authentic. Practically all records of the event draw heavily for details on the legend. What seems certain is that the event took place in his twenty-ninth year, soon after the birth of his son Rahula. Although Gautama must have been contemplating renunciation for a long time, the arrival of his son and heir served to settle the issue. No doubt he felt that if he did not decide to break away from his home immediately, he would merely continue to get more and more deeply entangled in the obligations of a householder and patriarch, which would make the final severance of family ties, if not impossible, at least much more difficult and painful. It was time to make a quick decision.

The earliest Buddhist accounts maintain that this decision was made on the very day of his son's birth. Gautama had gone for a walk by the riverside when the child was born. Gautama's father immediately sent a messenger to convey to him the glad tidings that his wife, Bhaddakaccana, had at last borne him a son and heir. But to the surprise of every one present Gautama did not go into transports of delight. Instead he looked distinctly crestfallen; and it was obvious from his general demeanour that he was not very pleased with the prospects of paternity. He went as far as expressing his dissatisfaction in so many words. 'A son is born,' he commented gloomily, 'a bond is born.' Disconsolate, his head bowed with the burden of this new responsibility, he returned home.

Meanwhile the news had been broadcast throughout the town of Kapilavastu, and the inhabitants were busy celebrating the birth of a grandson to their First Citizen. Large crowds had gathered at the gates of the city to congratulate Gautama, to acclaim him as a hero for having so effectively fulfilled his duty towards the State, his parents, his wife, and himself. While he was being taken through the streets in a triumphal procession, a certain maiden named Kisa Gotami, who had always secretly cherished the image of the Bodhisattva in her loving heart, on beholding 'the beauty and glory' of her hero from one of the roof-tops, could restrain her rapture no longer. She broke out into a loud song in his praise. 'Happy indeed is the mother, happy indeed is the father, happy indeed is the wife, who has such a husband.' Gautama heard her song, but it was the word 'happy' which seemed to echo in his consciousness like a mocking memory and intensified his misery. He could not be sure whether his mother, his father, and his wife were happy; but he was sure that he himself was far from happy. And how could the heart be happy? The eternal enigma tantalizingly presented itself to him. What was happiness? In his life as householder he had never experienced anything which could even remotely be described as a state of happiness. But doubtless there were other possibilities in life which he had yet to explore. And life was a brief candle. On this point he entertained no illusion. Dilatoriness was suicidal. It was time to make decisions; time to break loose from the ties of a life in which he had lost all faith and interest. The girl's simple song seemed to define for him the whole issue as though in a sudden flash of revelation. What is more, for no palpable reason at all, it seemed in a certain measure to strengthen his resolve regarding the renunciation he had so long contemplated. He felt a sense of gratitude towards Kisa for having thus unconsciously enabled him to make up his mind; and from his neck he took a pearl necklace which he sent to her, saying: 'Let this be your fee as a teacher.' But Kisa was not playing for such trifles as pearls, her heart was set on more precious stakes. Nevertheless, she was delighted with the gift. 'Young Siddhartha is falling in love with me,' she mused, 'and has sent me a present.' And she began weaving her dreams of further, more intimate favours. Unfortunately, however, these pleasant dreams were doomed to remain for ever unrealized, as Kisa herself must have discovered the next morning.

Towards the middle watch of that fateful night Gautama awoke to carry out his resolution. He saw his female musicians, who had fallen asleep round his couch, lying in postures of extreme abandonment, displaying their repulsive nakedness. The spectacle filled him with an intense aversion for a life dedicated to the pursuit of vulgar 'sense-desires.' He woke his charioteer Channa, and asked him to get his horse Kanthaka ready for him. The sleepy charioteer was unable to understand the object of this strange whim of his master's, but did as he was told. While Channa was gone to saddle the horse, the Bodhisattva felt a strong desire to take a last look at his wife and child. He went to her chamber, gently opened the door, and standing on the threshold for a few moments, watched in the dim light of an oil-lamp 'the mother of Rahula sleeping on her bed strewn with heaps of jessamine and other flowers . . . with her hand on her son's head.' He would have liked to hold the child in his arms, but he realized this might wake his wife and frustrate his whole design. He descended from his house to leave the town of his birth on his horse Kanthaka, with the miserable and sleepy-eyed Channa clinging to the horse's tail.

It was midsummer, and the warm depths of the earth tumescent under the silver caress of the full-moon. But the Bodhisattva's heart was turned against all 'sense-desires.' He rode through the night—and by daybreak he had travelled a safe distance from Kapilavastu. Then he revealed his intentions to Channa. At first the charioteer, who by now must have been completely exhausted, did not understand. However, when Gautama exchanged his clothes with a poor passer-by, cut off his long hair, and asked Channa to go back to Kapilavastu with the horse, the charioteer was 'overwhelmed with grief.' He began to weep. And not only the charioteer. The horse Kanthaka, too, says Asvaghosha, 'licked his feet with his tongue and dropped hot tears.'

Channa's grief was natural. He was very deeply attached to Gautama. Besides, he dreaded the thought of returning to Kapilavastu without his master; he knew he would be blamed for having helped Gautama in his plan of escape. He tried as best as he could to dissuade the foolhardy adventurer. 'Turn back, and have mercy on me,' his plea ended on a pathetic, personal note. Gautama was kind and consoling, but was not to be swayed from his resolve. Hopeless and tearful, Channa made a final appeal. 'Even if thy mind be resolved to abandon thy kindred and thy kingdom,' he begged, 'thou wilt not, O master, abandon me—thy feet are my only refuge.' But Gautama would neither agree to this. 'How will my father and my relations know what has become of me,' he argued with his weeping charioteer, 'unless you go back and tell them?' And after but a few more words Gautama went towards a hermitage in the neighbourhood.

Thus abandoned by his master, Channa 'tossed up his arms, wailed bitterly and fell on the ground' in his unavailing despair. Then, thoroughly sick in body and soul, the poor man turned back with the horse. The journey back to Kapilavastu was a sad affair for Channa. His progress was increasingly painful and slow; it took him full eight days to reach 'the city called after Kapila.' Presumably he must have wandered about the countryside, unable to muster up courage to face his master's reproachful family, and probably his mind was somewhat affected by the shock he had received; for Asvaghosha informs us: 'Sometimes he pondered, sometimes he lamented, sometimes he stumbled, and sometimes he fell; and so going along, wretched through his devoted attachment, he performed all kinds of actions in the road without conscious will.' At last, distracted and sorrowful, he managed to reach Kapilavastu. On his arrival there he soon discovered that his fears of what awaited him were not unfounded. The inhabitants of Kapilavastu, seeing him and the horse, thought at first that Gautama had also perhaps returned. But when he told them of what had happened, they were angry. Asvaghosha writes: 'Full of wrath, the people followed Channa in the road, crying behind him with tears, "Where is the king's son, the glory of his race and kingdom? He has been stolen away by thee." ' In vain, he protested that he had not kidnapped Gautama, but that Gautama had, of his own choice, abandoned the life of householder and gone into an uninhabited forest.

There was a veritable panic when Channa arrived at Suddhodana's house. Gautama's wife, who seemed to realize there was no hope of her husband's return, fainted, 'like the ruddy goose parted from her mate.' When she recovered from her swoon, she broke out in loud lamentations: 'He does not see that husband and wife are both consecrated in sacrifices . . .' she bewailed, mingling her sorrow with spite, 'Surely it must be that this fond lover of religion . . . has deserted me . . . in the hope to obtain heavenly nymphs in Indra's world.' However, realizing that it was no good wasting her anger on an absent husband, she turned her attack to Channa. 'Why dost thou weep to-day, O cruel one,' she shouted at him, 'having done a dishonourable, pitiless, and unfriendly deed to me? Cease thy tears and be content in thy heart—tears and that deed of thine ill agree.' She did not spare even the horse. 'The base creature,' she flung at the animal in her fury of frustration, 'now neighs loudly, filling the king's palace with the sound; but when he carried away my beloved, then this vilest of horses was dumb. If he had neighed and so awakened the people, or had even made a noise with his hoofs on the ground, or had made the loudest sound he could with his jaws, my grief would not have been so great.'

The horse was not in a position to protest. But Channa, his face down and his voice dim with sobs, tried to reason with her, pleading that neither he nor the horse were to blame for her lord's departure; that their sorrow was really as great as hers; and that if anybody must be held responsible it must be the gods. But 'a ruddy goose parted from her mate' could not be expected to listen to the voice of reason. Channa's ordeal was hard and long. Finally, when every one concerned had given vent to his emotions, and the tumult subsided, a family conference was called. After much discussion it was decided to send the family priest and a counsellor to Bhargava's hermitage, where Channa had left his master, and to persuade the impetuous youth to change his mind. But it was a fruitless journey. Those who once set out on a quest of the kind which Gautama had undertaken do not easily turn back.

VI

AFTER STRANGE GODS

FLESH and spirit are the two poles between which human longings eternally oscillate. At the very point where our longings for purely physical satisfaction begin to weary of their own ardours and seem almost to cease, there arise cravings for other, subtler forms of fulfilment. For at the root of things there is not only doubt, but desire—desire that is ever-restless because it is always unquenched and unquenchable. And so the process of oscillation goes on in an apparently endless recurrence. Boethius, therefore, cannot be regarded as the first man who turned to the consolations of philosophy when those less abstract had failed him tragically. Countless unhappy people before him had done the same. Gautama, for instance.

Even during his short career as householder, when he was trying conscientiously—if not successfully—to fulfil the exacting demands of family life, he had found time to dabble in the study of religion and philosophy. Indeed, the interest which he had shown in this direction was looked upon by his worldly-wise parents and kinsmen with marked disfavour. They all admired religion and philosophy, and honoured those engaged in the pursuit of spiritual ends—but they admired and honoured from a safe distance. It was entirely proper and right for the nobles and their sons to patronize the Order of Philosophers, pay homage to mystics and religious men, and to know enough about these matters to discuss them in polite company. But it was not considered proper to devote oneself to such barren, if exalted, pursuits in all seriousness; and, naturally, Gautama's family had done everything in their power to discourage this tendency in him when they realized that his metaphysical interests went beyond the limits of discretion.

However, undaunted by their disapproval, he had continued to follow his interests. In his imagination he had felt strongly drawn to the ascetic ideal. The figure of world-renouncer, unperturbed by the ebb and flow of humanity, and happy in his quiet solitude, had seemed extremely attractive to Gautama when seen in contrast to the ineffectual crowd in his immediate circle. Also, though so far his excursions into philosophy had been more in the nature of dilettante attempts, they were sufficient to convince him that, ultimately, a philosophic attitude afforded a far better safeguard against psychological turmoil and moral adversity than the pomp of wealth. And now that he had at last cast off the shackles of a conventional life, he felt himself free to take up his enquiry in earnest, and follow his vague intuitions to their logical end.

For a seeker after metaphysical consolations, India at the time was an ideal land. The country was seething with an intense spiritual ferment. Perhaps in no other part of the world, with the possible exception of Greece, had the people as yet achieved such an acute degree of intellectual self-consciousness. It is even arguable that the efflorescence of the speculative genius of the Indian people witnessed in Gautama's days has never been surpassed at any subsequent epoch of Indian civilization. At least, it is certain that the foundations of practically all the philosophical schools which flourished later in India were laid during this early period; and if the thinkers and metaphysicians of the time had not succeeded in evolving any proper systems, it was chiefly because their thought was at once too virile and fluid to admit of rigid systematization. Indeed, Gautama and his contemporary intellectuals were blessed in as much as they could explore countries of the mind unhampered by scholastic tyranny and academic mediocrity.

As ever, this outburst of intellectual activity had its origin in the heightening of man's interest in his own destiny and that of his world; his desire more fully to understand his own meaning and purpose, and the meaning and purpose of the infinitudes enveloping him; and his irrepressible urge to define anew the relation between himself and the universe. There was, most important of all, his craving for personal salvation; a craving, that is, to find some compensations for the deficiencies of his actual experience in the realms of fantasy.

To satisfy this demand of the human psyche, a great many devices had been elaborated, ranging from excruciating methods for self-mortification to Priapic rites of the most sensational nature. The religion practised by a large majority of the people was, of course, just a vast conglomeration of superstitions and magical beliefs derived from all kinds of sources, from lands lying as far apart as Egypt and the Central Asian Steppes, Mesopotamia and the country of the Middle Earth. In its totality the popular religion corresponded to a vague polytheistic animism, welded together by the common belief in ritualistic sacrifice as a means to the propitiation of deities, and in a hierarchy serving as the mediator between gods and mortals. But on higher levels of intelligence, need for rationalizing these beliefs had been felt. The religious sentiment had left its primitive simplicity far behind, had gradually grown in sophistication. The sense of 'the Holy' had become tempered with intellectual curiosity, if not actual scepticism. Ingenious schemes, like the Laws of Karma and Transmigration of Souls, had to be invented by the priests to strengthen their hierarchical institutions, which, it seems, had begun to record the first shocks from the side of reason as early as the eighth century B.C. Through these laws, which gave the priestcraft a strong theoretical sanction, while at the same time fixing moral principle on to an a-moral universe, the Brahmans made their own position impregnable for all practical purposes, not only in the immediate future but for a long long time to come.

Apparently, Gautama and other sceptics of his age had had anonymous forerunners several centuries earlier. The rudiments of scepticism are to be found in the Rig-Veda itself. 'Who is the God to whom we shall offer sacrifice?' An early heckler had harassed the pompous pundits of his time with this disquieting question, almost in the same spirit of mischief as a Hyde Park heretic might pester the Catholic Evidence lecturers with the query: 'Is the Pope a Christian?' And the mischief was done. Once the questioning had started there was no end to it. Dangerous lines of enquiry had suggested themselves; the human mind, with its hankering for the unknown and the hazardous, had taken them up, and in some instances pursued them beyond the limits laid down by experts in sacred matters. Was there only one God, or were there many gods? Questions and doubts had multiplied. Divisions had arisen among the Brahmans themselves. On the one side, there were worshippers of the Many; on the other, votaries of the One. There was a prolonged verbal feud between polytheists and monotheists, between pluralists and monists.

Eventually a happy compromise was reached, the dispute being settled by the clever admission that both parties were equally right. The conflict between the upholders of the One and the Many, said the arbiters, was altogether spurious. There was nothing fundamentally irreconcilable in both views; they were, in fact, complementary. The one and the Many were two sides of the same reality. In the beginning, they argued, there was a Unity; and then this Unity had of its own sweet will decided to break itself up into a Multiplicity. It was all very simple and straightforward!

The Brahmana of the Hundred Paths had even given an anthropomorphic description of the process. The One, moved partly by boredom and partly by a spirit of adventure, had wished: 'May I become a plurality—may I propagate myself.' The wish was followed by appropriate action: 'He exerted himself—he took on himself some pangs.' And 'when he had exerted himself, when he had endured some pangs, he created the Brahma.' Once Brahma had come into being, he begot and multiplied at an astonishing rate. Thus the plurality had come into operation. However, through this plurality there ran the thread of a yearning Spirit, the nostalgic Atman, which, by growing 'weary of wandering in a world of gloomy, formless phantasms,' slowly found its way back into the bosom of the One. In this manner, the One had become the Many; and yet had lost none of its uniqueness. A vast ocean had thrown out an infinity of ripples—and then started taking them back into its depths.

This was how the early Vedanists conceived of their Absolute. It was a great and hollow womb which could take almost anything into Itself, without ever being any the worse for it. The conception was extremely clever; and for a while it restored peace and good will among various rival schools of metaphysicians. Then, when everything seemed to be going smoothly, a mysterious individual caused panic by throwing a real bombshell in the metaphysical world. This enigmatic person was called Kapila, and he was known to have lived in Kapilavastu about a hundred years before Gautama's birth. Even his contemporaries knew little as to his actual life beyond the fact that, with a series of devastating aphorisms, he upset the admirably balanced apple-cart of the Absolute. The philosophers of the One, and of the Many, both dreaded facing him in argument. He had no use for either in their Vedantic form.

In his system, which came to be known as the Samkhya (literally, *numeral, rational or discriminative*), there was room only for a duality—the eternally co-existent duality of *Prakriti* and *Purusha*. The universe of which we know, said Kapila, is the product of the constant interaction of these two categories. There might be a supreme deity behind and beyond this duality; but as far as could be ascertained It played little part either in the origination or evolution of the universe. His doctrine was considered highly explosive and dangerous by the representatives of Brahmanical thought. And naturally enough. For though Kapila did not deny the existence of God in so many words, his system rendered an intelligent Creator superfluous in the scheme of things—and by implication, deprived the mediators between God and his creatures, or the priests, of their very *raison d'être*. This was serious heresy; and he was denounced in loud terms as an atheist, a propagator of false and godless doctrines. These charges, which he denied, were not altogether without truth.

Kapila had avoided taking up an openly atheistic position. But there soon followed in his wake proper, full-blooded atheists. The endless wrangles between conflicting philosophical schools had raised the important issue of clear definitions. To discuss on an intelligible level it was necessary to have a method or technique of argument; and a logic had actually been developed—a logic which was later on to become the basis of *Nyaya*, one of the six systems of Indian philosophy.

With the development of logic there had inevitably appeared logical realists. They would not be satisfied with vague postulates and mystical revelations; in argument they demanded substantial and concrete proofs. Some of them made no secret of the fact that they regarded immediate sensory experience as the only real source of knowledge. Thought, consciousness, and spirituality, they described as a property of matter, which 'alone is knowable and real.' They rejected the validity of the so-called intuitional knowledge, but at the same time they refused to recognize '*the course of reason*' or the idealist postulate. They could believe only in what they could see, feel, hear, touch or smell. And since Brahma could not be seen, heard, felt, touched or smelt, they were sceptical about his existence. They believed in the 'bloody horse'—and no more; Brahma, obviously, was not 'the bloody horse.' They argued: 'When any one says that "that is an ox, that is a horse" it is thereby pointed out. Point out to us the revealed Brahma, the Atman which dwells in everything.' The supporters of Brahma could only shrug their shoulders in despair faced by such radical scepticism.

However, there was another kind of realist who forced the Brahmans to more than merely shrugging their shoulders. These called themselves Lokayatikas, or the worldly-wise. But the Brahmans, in their spite, gave them the libellous designation of Charvakas—those who eat greedily. The Brahmans had good reasons to detest the Lokayatikas. The latter pressed their criticism of the Brahmanical point of view beyond the limits of religion, metaphysics, and logic. This the Brahmans considered hitting below the belt, and consequently very much resented it. For the Lokayatikas attacked the priests not only on intellectual or metaphysical but on moral grounds. They accused them not merely of being dialectical charlatans but also of hypocrisy and dishonesty. The Brahmanical system, said the Worldly-Wise, was not only logically false: it was a conscious and unscrupulous fraud:

'There is no heaven, no final liberation, nor any soul in another world,

'Nor do the actions of the four castes, orders, etc., produce any real effect.

'The Agnihotra, the three Vedas, the ascetic's three staves, and smearing one's self with ashes,

'Were made by Nature as the livelihood of those destitute of knowledge and manliness.

'If a beast slain in Jvotistoma [sacrificial ritual] rite will itself go to heaven,

'Why then does not the sacrificer forthwith offer his own father?

'If the Sraddha produces gratification to kings who are dead,

'Then, here, too, in the case of travellers when they start it is heedless to give provisions for the journey.

'If beings in heaven are gratified by our offering the Sraddha here,

'Then why not give the food down below to those who are standing on the housetop?

'While life remains let a man live happily, let him feed on ghee even though he runs in debt.

'When once the body becomes ashes, how can it ever return again?

'If he who departs from the body goes to another world,

'How is it that he comes not back again, restless for the love of his kindred?

'Hence it is only as a means of livelihood that Brahmans have established here

'All these ceremonies for the dead—there is no other fruit anywhere,

'All the well-known formulãe of the pundits, japhari, taphari, etc.

'And all the obscene rites for the queen commanded in the Asvamedha,

'These were invented by buffoons, and so all the various kinds of presents to the priests,

'While the eating of flesh was similarly commanded by night-prowling demons.'

In striking contrast to the flamboyant and rather bumptious self-confidence of the Worldly-Wise, there was the dark despondency of pessimistic fatalists like Makkhali Gosala, who denied all free-will and choice to man, and held out no hope or consolation for humanity. 'There is no power of action,' bemoaned Makkhali Gosala, 'man has no strength, man has no control; all beings, everything that breathes, everything that is, everything that has life is powerless, without power or ability to control [its own actions]; it is hurried on to its goal by fate, death, rebirth. Every being passes through a fixed series of rebirths, at the end of which the fool as well as the wise man comes to a final pause.'

Purana Kassapa, a kindred spirit, went one step further than Makkhali Gosala in hopelessness, and rejected the consolation of a moral law and its universal concomitant—the concept of reward. 'If a man,' he preached, 'makes a raid on the south bank of the Ganges, kills and lays waste and lets lay waste, burns and lets burn, he imparts no guilt to himself; there is no punishment of guilt. If a man crosses to the north bank of the Ganges, distributes and causes to be distributed charity, offers and causes to be offered sacrifices, he does not thereby perform a good work; there is no reward for good work.' Another indifferentist maintained: 'The wise man and the fool, when the body is dissolved, are subject to destruction and annihilation. Neither of them is beyond death.'

As the metaphysical questions had multiplied, so too had the doctrines which purported to answer them. The *Brahmajala-sutta*, or *The Discourse of The Net of Brahma*, enumerates as many as sixty-two different modes of speculation current among the metaphysicians of the period. The more important of these speculative schools were the eternalists, or those who on four grounds maintained that the soul and the world were eternal; the semi-eternalists, or those who maintained that the soul and the world were partly eternal and partly non-eternal; the out-and-out non-eternalists who regarded both the world and the soul as perishable goods; the extensionists, or those who conceived of the universe as a spatial reality, and in four ways set forth the infiniteness or finiteness of things; the fortuitous-originists who described the First Cause as having been an accident; the believers in a conscious after-life; the believers in an unconscious after-life; the non-believers in an after-life; the devoted theists and aggressive atheists.

As in Greece so in India, side by side with the growth of metaphysical and dialectical schools, there had arisen groups of sophists and sceptics who, presumably, were not particularly keen on arriving at the truth, and whose efforts were all directed to proving their opponents wrong rather than proving themselves right. They seemed to take a diabolical delight in pricking the speculative bubbles of solemn and serious contemporaries. But when they were called upon to set forth their own positive scheme of reality, they would immediately resort to some such equivocation: 'If you ask me whether there is another world—well, if I thought there was, I would say so. But I don't say so. And I don't think it is thus or thus. And I don't think it is otherwise. And I don't deny it. And I don't say there is, or is not, another world.'

Besides this passion for theological and metaphysical discussions—which then, as for a long time afterwards, served as a sort of national sport—there was a raging craze for all kinds of occult and magical practice among ignorant and ambitious souls. Mystics and magicians, quack doctors and sorcerers were legion, claiming huge followings. We frequently hear of Brahmans, 'who live on the food provided by the faithful, continue to gain a livelihood by such low arts, by such lying practices as these: that is to say, by divination from marks on the body; by auguries; by the interpretation of prognostics, of dreams, and of omens good or bad; by divinations from the manner in which cloth and other such things have been bitten by rats; by sacrifices to the god of fire, offerings of Dabba grass, offerings with a ladle, offerings of husks, of bran, of rice, of clarified butter, of oil, and of liquids ejected from the mouth; and by bloody sacrifices, by teaching spells for preserving the body, for determining lucky sites, for protecting fields, for luck in war, against ghosts and goblins, to secure good harvests, to cure snake bites, to serve as antidote for poison, and to cure bites of scorpions or rats; by divination, by the flight of hawks, or by the croaking of ravens; by guessing at the length of life, by teaching spells to ward off wounds; and by pretended knowledge of the language of beasts . . ; by explaining the good and bad points in jewels, sticks, garments, swords . . . women, men, youths, maidens, male and female slaves, elephants, horses, etc. . . ; by predicting future events . . ; by foretelling the eclipses of sun, moon, and planets . . ; by giving advice touching the taking in marriage or giving in marriage, the forming of alliances or the dissolution of connections . . ; by teaching spells to procure prosperity or to cause adversity to others, to remove sterility, to produce dumbness, locked-jaw, deformity, or deafness; by obtaining oracular responses by the aid of a mirror, or from a young girl, or from a god; by worshipping the sun, or by worshipping Brahma; by spitting fire out of their mouths, or by laying hands on people's heads . . ; by teaching the ritual for making vows and performing them, for blessing fields, for imparting virility and rendering impotent . . ; by prescribing medicines to produce vomiting or purging, or to remove obstructions in the higher or lower intestines, or to relieve headaches; by preparing oils for the ear, collyriums, catholicons, antimony, and cooling drinks; by practising cautery, midwifery, or the use of root-decoctions or salves.' Apparently, these quacks knew the value of advertisement. They did not believe in being humble about their talents. 'I know no Samana, no Brahman, no teacher, no master, no head of the school, even though he calls himself the holy supreme Buddha,' proclaims a certain swaggering spiritual healer named Saccaka, 'who, if he faces me in debate, would not totter, tremble, quake, and from whom the sweat would not exude. And if I attacked a lifeless pillar with my language, it would totter, tremble, quake—how much more then would a human being.'

There were also a large number of whimsical holies. The vow taken by some of them consisted in bringing their behaviour as closely into line with that of a cow or a hen as was humanly possible; they would go about picking their food from the ground in the manner of a hen, or graze in the fields in imitation of cows, and were, consequently, known as 'hen' or 'cow' saints. Others believed in the purifying efficacy of water to such an extent that, in order to wash away all impurities of the flesh no less than those of the spirit, they spent all their time in purging themselves by drinking large quantities of water. There were ascetics, especially among the followers of Mahavira the Jäin leader, and Makkhali Gosala the fatalist, who undertook extremely agonizing austerities. Their methods of self-mortification included such ingenious devices as prolonged fasts, and in some extreme cases even starvation to death, remaining poised on their head or on one leg for long periods, lying on beds of thorns, not washing themselves and eating filth. And there were, of course, quite a large number of world-renouncers whom the Greeks later called gymnosophists, who considered themselves absolutely 'free from worldly fetters,' and so dispensed with even their clothes, and went about shocking the public by the unashamed display of their nakedness.

It would be an impossible task to set out to give an exhaustive list of all the doctrines, beliefs, and practices current at the time when Gautama embarked on his spiritual adventure. Those enumerated above should give some vague idea of the prevailing intellectual confusion. It would be a mistake to regard this confusion of ideas and faiths as a symptom of the spiritual degeneracy of the age. On the contrary, it reflected a keen and healthy spirit of intellectual experiment and adventure, a vigorous zest for things of the mind, and a surplus of emotional energy, even though it may be true that these enthusiasms sometimes yielded results which were not entirely free from ludicrousness.

Gautama's own experiments in search of truth were of a more serious character. After leaving Channa, his charioteer, he made his way to Rajagaha, the capital of Magadha, which was a flourishing city, situated in an idyllic valley engirdled by five hills, and renowned for its palaces. It was the seat of Bimbisara, one of the most influential potentates in the eastern valley of the Ganges. And king Bimbisara, though he was a merry monarch, managed to combine his passion for beautiful courtezans with an equally live interest in metaphysics. As a result, a great many hermit philosophers had settled down in the caves on the hillsides in the neighbourhood.

Gautama had gone to Rajagaha with the purpose of studying the doctrine and discipline of Alara Kalama, about whom he had probably heard while still living in Kapilavastu. Alara's doctrine drew largely on the metaphysics of Kapila, the founder of the Samkhya school. It divided the world into two distinct categories of 'the evolvent' and 'the evolute,' 'the manifested' and 'the unmanifested.' Its *summum bonum* was represented by a quasi-nihilistic freedom from the chain of causation, or the achievement of a state of Nothingness.

Alara readily accepted Gautama as a disciple, explained to him the tenets of his doctrine, and taught him what he thought to be the path of deliverance. For some time Gautama tried to follow this doctrine. But the results were not very satisfactory and eventually he abandoned it, 'because this doctrine extending to the Attainment of the state of Nothingness did not conduce to aversion, absence of passion, cessation, tranquillity, higher knowledge, Nirvana.' He then studied for a while under another teacher, Uddaka Ramaputta. Uddaka was not an absolute nihilist, but stopped short at an intermediate state of Neither-consciousness-nor-nonconsciousness. We are told that 'having learned the inherent imperfections of the name and the thing named, he took refuge in a theory beyond Nihilism, which maintained a name and a non-name'; and further 'since even a name and a non-name were substrata, however subtle, he went . . . and found his restlessness set at rest in the idea that there is no named and un-named.' As far as Gautama was concerned, he found the state of Neither-consciousness-nor-nonconsciousness of Uddaka as sterile as the state of Nothingness held up as a bait by his previous master, and he abandoned it after a time.

It is quite possible that in addition to serving his spiritual apprenticeship under Alara and Uddaka, Gautama experimented with the semi-mystical, semi-metaphysical systems of many other contemporary teachers. But the next really important event in his life was his undertaking severe austerities. From times immemorial in practically every religion people have, in one form or another, believed that through bodily penance and, self-mortification man can achieve supernatural insight, and realize a state of bliss beyond the sphere of pleasures he can receive through the antennae of his senses. This belief has probably been held more widely in India than anywhere else. No doubt Gautama wanted to determine for himself the truth behind this popular belief. In a dialogue describing his experiments with truth, he says:

'Then striving after the good, and searching for the supreme state of peace, I gradually made my way to the Magadhas, and went to Uruvela, the army township. There I saw a delightful spot with a pleasant grove, a river flowing delightfully with clear water and good fords, and round about a place for alms . . .'

He decided that it was an ideal place for 'one intent on striving'; and so he began his penance, attended by five disciples. His discourse gives a very faithful account of his exertions:

'Then I thought, what if I now set my teeth, press my tongue to my palate, and restrain, crush, and burn out my mind with my mind. (I did so) and sweat flowed from my armpits. Just as if a strong man were to seize a weaker man by the head or shoulder . . . so did I set my teeth . . . and sweat flowed from my armpits. I undertook resolute effort, unconfused mindfulness was set up, but my body was unquiet and uncalmed, even through the painful striving that overwhelmed me. Nevertheless such painful feeling as arose did not overpower my mind.

'Then I thought, what if I now practise trance without breathing. So I restrained breathing in and out from mouth and nose. And as I did so, there was a violent sound of winds issuing from my ears. Just as there is a violent sound from the blowing of a blacksmith's bellows, even so as I did so there was a violent sound . . . Then I thought, what if I now practise trance without breathing. So I restrained breathing in and out from mouth, nose, and ears. And as I did so violent winds disturbed my head. Just as if a strong man were to crush one's head with the point of a sword, even so did violent winds disturb my head . . .'

He practised holding his breath again three times, and the pains were 'as if a strap were being twisted round his head, as if a butcher were cutting his body with a sharp knife, and as if two strong men were holding a weaker one over a fire of coals.' Having completed his respiratory yogic exercises, he moved to the next stage of his penance which consisted in benumbing the senses through gradual starvation. The discourse continues and gives a vividly realistic picture of this terrible form of self-immolation:

'Then I thought, what if I were to take food only in small amounts, as much as my hollowed palm would hold, juice of beans, vetches, chickpeas, or pulse. (I did so.) My body became extremely lean . . . The mark of my seat was like a camel's footprint through the little food. The bones of my spine when bent and straightened were like a row of spindles through the little food. As the beams of an old shed stick out, so did my ribs stick out through the little food. And as in a deep well the deep low-lying sparkling of the waters is seen, so in my sockets was seen the deep low-lying sparkling of my eyes. And as a bitter gourd cut off raw is cracked and withered through wind and sun, so was the skin of my head withered through the little food. When I thought I would touch the skin of my stomach, I actually took hold of my spine, and when I thought I would touch my spine, I took hold of the skin of my stomach, so much did the skin of my stomach cling to my spine through the little food. When I thought I would ease myself, I thereupon fell prone through the little food. To relieve my body I stroked my limbs with my hand, and as I did so the decayed hairs fell from my body through the little food.

'Some human beings seeing me then said, "The ascetic Gautama is black." Some said, "Not black is ascetic Gautama, he is brown." Others said, "Not black is the ascetic Gautama, nor brown, his skin is that of a Mangura fish [a kind of sheat-fish], so much had the pure clean colour of my skin been destroyed by the little food.

'Then I thought, those ascetics and brahmans in the past, who have suffered sudden, sharp, keen, severe pains, at the most have not suffered more than this . . . But by this severe mortification I do not attain superhuman truly noble knowledge and insight. Perhaps there is another way to enlightenment . . .

'Then I thought, it is not easy to gain that happy state while my body is so very lean. What if I now take solid food, rice, and sour milk . . .'

The thought was followed by the appropriate act. But Gautama's decision to take solid food, because the shores of the Great Beyond could not be reached on an empty stomach and an emaciated body, greatly shocked the five monks who had been watching his agony in the fervent hope that, after having gained the Doctrine, Gautama would tell it to them. His reverting to a normal diet of 'solid food, rice, and sour milk,' they looked upon as a personal affront which could not be condoned. And in a sense it was: it destroyed all their chances of discovering the Doctrine by proxy. They left him in disgust, saying: 'The ascetic Gautama lives in abundance, he has given up striving, and has turned to a life of abundance.'

Between Gautama's renunciation and his enlightenment

there elapsed six years. He was twenty-nine when he left home; thirty-five when he gained wisdom. It was a long quest. In the course of his search he wandered from place to place, from teacher to teacher. He had tried to master several doctrines but found them wanting in the thing he was striving for; he had worshipped at many strange altars but found no solace. He had subjected himself to the most painful penance of which any man is capable—and discovered there was nothing in it, nothing at all. He had been almost at death's door. His most devoted companions had left him in the moment of his greatest need. Even the strong bonds of friendship, he had seen, were an illusion. And at the end of all this struggle and suffering, he appeared to have achieved nothing. The quest for a land of heart's desire had proved a failure. So it seemed to all those who had watched his career with interest; so, too, perhaps it seemed to his own mind in moments of despair.

And yet, perhaps, it had not been entirely fruitless. It had defined his essential problems. The issues about which he had been quite vague in the beginning were now clear. And it had brought him at least one realization of paramount importance: the realization that if the land of heart's desire could not be reached by any means known to man, then it might very well be because there was no such land. Without this realization Gautama could never have attained enlightenment; for it seems indeed to be of the very nature of that enlightenment.

VII

ENLIGHTENMENT

> The Tathagata, brethren . . . he it is who doth cause a way to arise which had not arisen before; who doth bring about a way not brought about before; who is the knower of a way, who understands a way, who is skilled in a way . . .
>
> *Samyutta-Nikaya*

IT is by no means an established fact that enlightenment came to Gautama while sitting cross-legged under the bodhi-tree on eight handfuls of grass which a grass-cutter, Sotthiya, had given him, and which, when he had shaken it out, had assumed the shape of a comfortable seat 'fourteen hands long.' The incidental circumstances attending the enlightenment are, of course, quite extrinsic to the real issue; for in any case it matters little how and where Gautama received his initiation of wisdom. Even if we did know that the event took place under a bodhi-tree, a rose-apple tree, or a bamboo tree, we should be none the wiser as to its real meaning. However, in passing it may be pointed out that one of the very early accounts does not mention the famous bodhi-tree (*ficus religiosa*) at all, and completely ignores the obliging grass-cutter and the folding grass-seat.

The question to be decided with regard to Gautama's enlightenment is to discover what precisely it did signify. It is not an easy question; indeed, in a very large measure, it is insoluble. The true nature of Gautama's experience must for ever remain a mystery to us. He alone knew what it meant, and he said very little about it. Whatever evidence there may be, it is too scanty for any dogmatic assertions. The utmost we can do is to offer a tentative interpretation and indicate a possible line of understanding. Naturally, in a case like this, a number of interpretations are possible; and all of these may be equally true or untrue. We have to choose our own point of emphasis, and this choice cannot help being more or less personal, and will inevitably be determined by our particular predilections.

As far as can be seen, from the earliest times, hagiographers have been inclined to interpret Gautama's enlightenment in two distinct ways. On the lower level there is the melodramatic and miraculous view. It represents the process through which the enlightenment was reached as an almost physical process of conquest of supernatural power; the dramatic effect being realized by staging a full-blooded battle between Gautama and the forces of evil. This view finds its crudest and most exciting expression in the legend of Gautama's combat with Mara, the tempter. The legend is to be found both in the Pali and Sanskrit sources, though there are a number of variants of the temptation-theme which differ from one another in minor details. The whole story of the assault of Mara is too fantastic to bear repetition; but to show some idea of the lines on which the experience is unfolded we might give the description of these cosmic upheavals which apparently preceded the actual combat between the Bodhisattva and the 'Evil One':

'When the conflict began between the Saviour of the world and the Prince of Evil a thousand appalling meteors fell; clouds and darkness prevailed. Even this earth, with the oceans and mountains it contains, though it is unconscious, quaked like a conscious being—like a fond bride when forcibly torn from her bridegroom—like the festoons of a vine shaking under the blasts of a whirlwind. The ocean rose under the vibration of a whirlwind. The ocean rose under the vibration of this earthquake; rivers flowed back towards their sources; peaks of lofty mountains, where countless trees had grown for ages, rolled crumbling to the earth; a fierce storm howled all around; the roar of concussion became terrific; the very sun enveloped itself in awful darkness, and a host of headless spirits filled the air.'

Rhys Davids saw in the story of 'a visible Tempter' an attempt to objectify Gautama's inner conflict after he had given up the penance, and all his old temptations had come back upon him with renewed force. 'In the legend,' he wrote, 'the very thoughts passing through the mind of Gautama appear in gorgeous descriptions as angels of darkness or of light . . . but they have still a depth of meaning to those who strive to read between the lines of these, the first half-inarticulate efforts the Indian mind had made to describe the feelings of a strong man torn by contending impulses.' He also emphasized curious points of resemblance between Milton and the Buddhist poets; and, indeed, the resemblance is unmistakable. There is a great deal in the surcharged imagery of the Buddhist poets—an imagery which often borders on the grotesque—which reminds one of Milton's grandiloquent outbursts. The description of the cosmic convulsion given above, and several passages in Asvaghosha's version of the assault of Mara, for instance, might well be paralleled by passages from *Paradise Regained*, like the following:

. . . And either tropic now
'Gan thunder, and both ends of heaven; the clouds
From many a horrid rift abortive pour'd
Fierce rain with lightning mix'd, water with fire
In ruin reconciled; nor slept the winds
Within their stony caves, but rushed abroad
From the four hinges of the world, and fell
On the vex'd wilderness, whose tallest pines
Tho' rooted deep as high, and sturdiest oaks,

BUDDHA IN SHRINE
(Ajanta—*The Director of Archaeology, Hyderabad*)

Bow'd their stiff necks, loaden with stormy blasts,
Or torn up sheer. Ill wast thou shrouded then,
O patient Son of God, yet only stood'st
Unshaken! Nor yet staid the terror there:
Infernal ghosts and hellish furies round
Environ'd thee; some howl'd, some yell'd, some
shriek'd,
Some bent at thee their fiery darts, while thou
Sat'st unappall'd in calm and sinless peace.

But it is very doubtful whether the Buddhist poets had any conscious intention of communicating 'spiritual truths' by means of an imagery drawn from tangible things. On the contrary, it would seem that their main reason for resorting to this uncouth symbolism was, that they were unable to understand Gautama's crucial experience in any other terms. Rhys Davids himself admitted as much when he said: 'It may be questioned how far the later Buddhists have been able to realize the spiritual truth hidden under these material images; most of them have doubtless believed in a real material combat, and a real material earthquake.' And in this respect there is no reason to suppose that the early Buddhists possessed any greater capacity for psychological subtlety than the later ones.

Of course, both in the early and late periods, there were individual Buddhists to whom Gautama's enlightenment signified something totally different from a physical feat. And because they saw it as something of a 'spiritual' order, they tried to convey its meaning in a totally different language. What they give us is not a man fighting with demons of darkness, but a mystic struggling to realize in himself that inner vision which can penetrate the veil of temporal relationships and see what is behind and beyond. They have actually described the whole process through which Gautama discovered the right mode of meditation, and attained the mystical *élan*—a state of continuous Present, of poise and harmony, which has no place in it for the contrarieties and distinctions which are the lot of ordinary mortals. With an almost scientific precision they outline the curve of Gautama's experience, and show him gradually eliminating all the avenues of mental and sensory knowledge in order to rise to that condition of super-consciousness where knowledge is immediate, intuitive, and final. The account of it is given in the *Mahasaccaka-sutta*. It is written in the first person singular, but that is no guarantee that it is an accurate report of Gautama's own words:

'Now having taken solid food and gained strength, without sensual desires, without evil ideas I attained and abode in the first trance of joy and pleasure, arising from seclusion and combined with reasoning and investigation. Nevertheless such pleasant feelings as arose did not overpower my mind. With the ceasing of reasoning and investigation I attained and abode in the second trance of joy and pleasure arising from concentration, with internal serenity and fixing of the mind on one point without reasoning and investigation. Nevertheless such pleasant feeling as arose did not overpower my mind. With equanimity towards joy and aversion I abode mindful and conscious, and experienced bodily pleasure, what the noble ones describe as "dwelling with equanimity, mindful, and happily," and attained and abode in the third trance. Nevertheless such pleasant feeling as arose did not overpower my mind. Abandoning pleasure and abandoning pain, even before the disappearance of elation and depression, I attained and abode in the fourth trance, which is without pleasure and pain, and with purity of mindfulness and equanimity. Nevertheless such pleasant feeling as arose did not overpower my mind.

'Thus with mind concentrated, purified, cleansed, spotless, with the defilements gone, supple, dexterous, firm, and impassible, I directed my mind to the knowledge of the remembrance of my former existences. I remembered many former existences, such as, one birth, two births, three, four, five, ten, twenty, thirty, forty, fifty, a hundred, a thousand, a hundred thousand births; many cycles of dissolution of the universe, many cycles of its evolution, many of its dissolution and evolution; there I was of such and such a name, clan, colour, livelihood, such pleasure and pain did I suffer, and such was the end of my life. Passing away thence I was born elsewhere. There too I was of such a name, clan, colour, livelihood, such pleasure and pain did I suffer, and such was the end of my life. Passing away thence I was born here. Thus do I remember my many former existences with their special modes and details. This was the first knowledge that I gained in the first watch of the night. Ignorance was dispelled, knowledge arose. Darkness was dispelled, light arose. So is it with him who abides vigilant, strenuous, and resolute.

'Thus with mind concentrated, purified, cleansed, spotless, with defilements gone, supple, dexterous, firm, and impassible, I directed my mind to the passing away and rebirth of beings. With divine, purified, superhuman vision I saw beings passing away and being reborn, low and high, of good and bad colour, in happy or miserable existences according to their Karma. Those beings who led evil lives in deed, word, or thought, who speak evil of the noble ones, of false views, who acquire Karma through their false views, at the dissolution of the body after death are reborn in a state of misery and suffering in hell. But those beings who lead good lives, in deed, word, and thought, who speak no evil of the noble ones, of right views, who acquire Karma through their right views, at the dissolution of the body after death are reborn in a happy state in the world of heaven . . . This was the second knowledge that I gained in the second watch of the night . . .

'Thus with mind concentrated, purified, cleansed, spotless, with the defilements gone, supple, dexterous, firm, and impassible, I directed my mind to the knowledge of the destruction of *asavas* [cankers]. I duly realized [the truth] "this is pain," I duly realized [the truth] "this is the cause of pain," I duly realized [the truth] "this is the way that leads to the destruction of pain." I duly realized "these are the *asavas*" . . . "this is the cause of *asavas*" . . . "this is the way that leads to the destruction of *asavas*." As I thus knew and thus perceived, my mind was emancipated from the *asava* of sensual desire, from the *asava* of desire for existence, and from the *asava* of ignorance. And in me emancipated arose the knowledge of my emancipation. I realized that destroyed is rebirth, the religious life has been led, done is what was to be done, there is nought (for me) beyond this world. This was the third knowledge that I gained in the last watch of the night. Ignorance was dispelled, knowledge arose. Darkness was dispelled, light arose. So is it with him who abides vigilant, strenuous, and resolute.'

There is no need to emphasize the metaphysical and mystical character of this analysis. As it is presented here, the interpretation is comparatively simple and even naïve. However, it was from such humble beginnings that there was to spring up the vast and impressive edifice of Mahayana Metaphysics. The theory of a causal chain and the scheme of 'dependent origination,' which is dimly anticipated here, when followed to its logical conclusion was to lead to the Madhyamika doctrine of Sunyata—the *ne plus ultra* of all Critiques of Pure Reason. Certain quite innocent hints dropped in this interpretation were to encourage audacious metaphysicians like Nagarjuna to venture on a most hazardous flight through the limitless realms of 'absolute vacuity.' M. René Grousset has described this flight in words which cannot be improved upon. 'The sage,' he says, 'has gone down into the depths of his heart. He has seen there . . . the external world, in the phenomenon of representation, taking shape and vanishing there. He has seen the dissolution of all that we call ego, of the substantial soul . . . and of the phenomenal ego . . . In place of this world of moral suffering and material obstacles, of internal egotism and objective adversity, an apparently bottomless gulf opens in the heart—a luminous and as it were submarine gulf, unfathomable, full of ineffable beauties, of fleeting depths and infinite transparencies. On the surface of this vacuity into which the eye plunges dazzled, the mirage of things plays in changing colours, but these things, as we know, "exist only as such"—*tathata*—and therefore are as if they were not. And once this mirage is dispelled, behold—in the intimate contemplation of that bottomless and limitless depth, in that unrivalled purity of the absolute vacuity—behold all virtualities arising, all powers emerging.'

This is the other, more refined, interpretation of Gautama's enlightenment.

Both these interpretations have their modern exponents; and some typical instances have been anatomized elsewhere. There are even to-day, on the one hand, those who champion the concept of a 'fight' and on the other, those who repose their faith in the idea of a mystico-metaphysical flight through the boundless void. The first view has a powerful appeal for the mythic imagination. In fact, it is the mythological view; and writers like H. Kern are probably quite justified in suggesting that most of the legends of Buddha, and in particular the legend of his combat with Mara, the spirit of darkness, have their basis in the ancient astronomical lore, and personify the sun-myth. The latter view, because of its mystical implications, serves to furnish the spiritual élite with a sanction for their Credo. Taken together the two views complement each other.

And yet both views fail to be convincing. They fail to be convincing because if either of them were accepted Gautama's whole behaviour after his enlightenment becomes incomprehensible. The first view is in any case too infantile to be taken seriously. The second view cannot be rejected so easily. It has behind it the sanction of most eminent and capable authorities, modern as well as ancient. Nevertheless, in the ultimate analysis, it turns out to be as unsatisfying as the more popular interpretation. There is, in fact, sufficient warrant for rejecting it; and this warrant is furnished by the actual life led by Gautama from the time of his enlightenment right up to the day of his death.

By their deeds shall ye know them. Wherever there is doubt, a man's deeds must be regarded as the acid test. This is the criterion that must be applied to Gautama. He was thirty-five when he received enlightenment; he died when he was past eighty. What was he doing during the long period which separates these two events? The question is well worth considering. One thing is certain: he did not devote the rest of his life to 'an intimate contemplation of that bottomless and limitless depth' from which 'all virtualities arise, all powers emerge.' A preoccupation of this nature is not without certain fascination. The fact that a great many Mahayanist metaphysicians were positively enamoured of 'the unrivalled purity of the absolute vacuity' proves that it must have powerful attractions. Gautama, however, seems to have resisted its temptations quite as successfully as those of his dancing girls. He had apparently no use for 'the unrivalled purity of absolute vacuity.' His tastes were altogether more solid and less abstract. He was a matter-of-fact person, and after his visit to Alara and

Uddaka he had been confirmed in his conviction that it was a sterile occupation to spend all one's time in trying to attain the state of Nothingness, or the state of Neither-consciousness-nor-nonconsciousness. The years of his ministry he spent instead in attempting an alleviation of the lot of his fellow men, often at the cost of considerable inconvenience to himself. He travelled many a weary distance, and travelling in those days was anything but a pleasurable pastime. For more than forty years he wandered from village to village preaching a doctrine which was not at all esoteric, but as the late Sir Charles Eliot observed, 'essentially practical, human, business-like.' The people to whom he preached were frequently obstinate and stupid; in reasoning with them he had to have infinite patience and forbearance. He had often to suffer fools gladly. He was misunderstood, misrepresented—in some instances, openly maligned. His struggle against the respectable prejudices and superstitions of his age brought on him the abuse of vested interests—an abuse, which, says a Buddhist dialogue, 'was copious, not at all stinted.' What was the motive which induced him to return from the shores of the Great Beyond to undertake a thankless, and often hopeless, mission? Having reached the verge of 'the submarine gulf' of the Madhyamika school, he could easily have remained there to enjoy the beatific vision of its 'ineffable beauties, fleeting depths and infinite transparencies.' But as far as can be judged from his behaviour, he preferred attending to mundane affairs to the contemplation of mystic beatitudes. It is true, he insisted upon an objective view of reality, and a clear understanding of man and his place in the universe. But such a view and understanding were not for him ends in themselves; they were only means to an end. He advocated them because he regarded them as the indispensable precondition of a reasonable and decent way of living. And it was on this way of living that he placed the whole emphasis throughout the years of his ministry.

Gautama, indeed, never became a perfect *Sanyasi*. He never became a perfect *Sanyasi* because, in his unsentimental and detached way, he cared far too much for this world of transient things. Mrs Rhys Davids quotes an outspoken Indian Yogi who, 'when praised for furthering the welfare of others is supposed to have broken into a laugh and said: "What have I to do with the welfare of others? It takes me all my time to mind my own welfare!" ' This profound statement truly reveals the attitude of a perfect Yogi. However, Gautama could never share such divine indifference to the fate of humanity. He performed, it is true, no spectacular miracles. There was nothing of the wonder-working Messiah about him. Still less was he an ostentatious and sentimental philanthropist. But he was very far from being indifferent to the 'welfare of others.' He made the welfare of others very much his business; and it would be difficult to find a more inspiring example of selfless devotion to the cause of humanity. With his own hands he tended the sick and poor. Those who came to him to be comforted —and they were legion—found in him a kind and understanding friend. His tenderness and compassion evoked a sympathetic response even among those who considered his teachings to be a menace to social stability. He was always impressing upon his audience that in this 'ocean of wayfaring' which is life, human beings should lend each other a helping hand—because if they did not, there was no higher power which would help them in solving their problems. It was for this reason that he founded the Order. He intended it to be a voluntary association of dedicated persons who would devote themselves to the task of making the process of 'wayfaring' easier for such among their fellow beings as were weak, helpless, and stricken. It is another matter that the Order never quite became what it was meant to be.

There is also the warrant of his word. In fact, there is hardly any inconsistency between what he said and what he did. He was, of all the historical personages of whom we possess any knowledge, one of the most consistent in thought, word, and act. Louis de la Vallée Poussin calls Gautama '*le grand religieux*,' and suggests that the quintessence of his system, or '*le fait bouddhique*' as he defines it, is '*une certaine forme de Yoga ou ascétisme-mysticisme*.' Yet the surprising thing is that '*le grand religieux*' himself does not appear to have been at all anxious to lay any stress on this '*fait bouddhique*.' On the contrary, he said singularly little about his '*ascétisme-mysticisme*,' and deliberately discouraged any tendency on the part of his disciples to be obsessed with metaphysical and mystical issues. He not only placed little value on the supra-rational knowledge and ecstasy to which ascetics and mystics were supposed to have access, but actually described their mental acrobatics as 'the thicket of theorizing, the wilderness of theorizing, the tangle, the bondage and shackles of theorizing . . .'

In another discourse, which is significantly entitled *Of The Irrelevant*, he made his position in this matter still more clear. One of his disciples, Malunkya, it is said, had been feeling very much worried about some metaphysical problems. Unable to keep his doubts to himself any longer, he made up his mind one day to question Gautama point-blank on these issues. So, going up to him, he asked disarmingly: 'If the lord knows that the world is eternal, let him tell me so. If the lord knows this world is not eternal, let him tell me so. If the lord does not know whether the world is eternal or not, then the only straightforward thing for one who knows not, nor discerns, is to avow that he knows not nor discerns.'

Faced with this disconcerting enquiry Gautama asked Malunkya whether 'the lord' had made any promise to enlighten him on these abstruse matters as a condition of his joining the Brotherhood. Half-heartedly, Malunkya admitted that no such promise had been given. 'It comes to this then,' said Gautama, 'that I never promised, nor did you stipulate that, as a condition of your following the higher life under me, I should expound these matters to you.' And he went on to explain, by means of a parable, where precisely the emphasis should be and where it should not be. 'If a man were to say . . .' he argued, 'that he would not follow the higher life under me, until I had answered all the questions you enumerate, he would get no answer from me until death overtook him. It is as if a man were transfixed by an arrow heavily coated with poison, and his friends and kinsfolk were to get a leech-expert in dealing with arrow-wounds, but the man were to declare he would not have the arrow taken out until he knew whether the archer who had shot him was noble, or a Brahman, or a middle-class man or a peasant—what the archer's name was—whether he was tall or short or of a medium height—whether he was black or brown or fair—what particular village or township or city he hailed from—what kind of bow he had—whether his bow-string was made from bamboo, or hemp, or leaves of the *calotropis gigantea*—whether the shaft of the arrow was a wild reed or a planted shoot—whether the shaft was feathered with plumage of a vulture, or a heron, or a falcon, or a peacock, or some other fowl—whether the gut binding that shaft came from an ox, or a buffalo, or a hart, or a monkey—whether his arrow was a plain arrow, or was barbed with horn, or iron, or a calf's tooth, or with an oleander thorn. The man would never get to know this before death overtook him . . . The higher life,' he added, 'is not contingent on the truth of any thesis that the world is either eternal or non-eternal. In either case, as in each of other theses you adduce, there still abides the fact of birth, decay, and death; there still abide the facts of grief and tribulation, of ill, sorrow, and distraction—and of their extirpation in the here and now . . . I have not taught that the world is either eternal or not eternal; that it is finite or infinite; that life and the body are either identical or distinct; that after death a truth-finder passes or does not pass to a further existence, or does both or neither. And why have I left these things untaught?—Because they are unprofitable, not fundamental to the higher life of which I speak . . .'

All these factors tend to lead to a radically different interpretation of the nature of Gautama's experience. In the first place, it is quite clear that all along Gautama had followed the strictly scientific method of elimination through 'trial and error.' He had gone as far as it was possible with the metaphysical enquiry only to find that theoretical knowledge did not touch the heart of the matter, and that in fact it led one into a mental wilderness. With utmost scrupulousness he had practised the disciplines enjoined by well-known mystics and ascetics, even subjected himself to such horrible penances as wearing haircloth, and lying among decomposing corpses in the cemetery, but found them fruitless. With the realization of the futility of these methods there had been born in him an entirely new definition of the problem. The emphasis had shifted from the metaphysical to the moral issue. The question was no longer the attainment of a supra-conscious state of ecstatic bliss, but 'the extirpation of grief and tribulation, of ill, and sorrow, and distraction in the here and now.' This transference of emphasis was one aspect of enlightenment. The other was the working out of a technique of living which would be conducive 'to passionlessness . . . to tranquillity, to insight . . . and to Nirvana'—or, in other words, 'to the extirpation of grief and tribulation, of ill, and sorrow, and distraction in the here and now.'

There was thus no mystery attached to Gautama's enlightenment. It was a simple and intelligible experience; or, perhaps, it would be more correct to say that it was neither more nor less mysterious than, for instance, the discovery and formulation of the Laws of Gravitation. Gautama and Newton, it is true, worked in very different fields. Ultimately, however, that difference counts for little. What is more significant is the essential identity of their methods. Gautama's field of investigation was the universe of human mind, the relationships and laws that govern it, and the properties and behaviour of those subtle elements which enter into its structure. It is by no means an easy field to investigate on scientific lines. It is easy enough to be objective about a dead frog, but not so easy to be objective about one's own heart, pulsating with a whole world of desires. Nevertheless, Gautama did succeed in subjecting 'the world within' to some kind of critical scrutiny, and thus managed to introduce a certain measure of objectivity into the subjective chaos. This was his distinctive achievement; this, too, the meaning of his enlightenment.

The discovery of profoundest truths often comes from the recognition of simplest relationships. Consequently, in all such discoveries, there is an element of surprise, of instantaneousness, and even of mystery; that is to say, although one may have followed a long process of intellectual argument and experimental analysis, the final crystallization of truth seems to be effected in an unexpected and sudden manner—in a flash of revelation, as it were. Gautama's experience appears to have been of this character. As H. G. Wells rightly interprets it: 'When the mind grapples with a great and intricate problem, it makes its advances, it secures its positions step by step, with but little realization of the gains it has made, until suddenly with an effect of abrupt illumination, it realizes its victory. So it would appear to have happened to Gautama.'

The emphasis on 'the way' in *Samyutta-Nikaya* is not altogether without meaning: enlightenment was essentially the working out of a way of life that would bring about 'the extirpation of grief and tribulation, of ill, and sorrow, and distraction in the here and now.' In the very nature of things, this could not but be 'difficult to perceive, hard to realize.' It was a moral way; though 'moral' in a peculiar psychological sense. The rest of his life Gautama devoted to preaching this way. 'By their deeds shall ye know them'—and it is reasonable to assume that he was not so much interested in 'explaining the world' as—to borrow a phrase from Marx's criticism of Feuerbach—in 'changing it.' But Gautama was aware, as no doubt Marx in his own way was aware, that before attempting to change the world one must understand it 'such as it is.' In a later discourse Gautama was to declare: 'So long, monks, as I did not comprehend, as it really is, the satisfaction in the world as such, the misery in the world as such, the escape therefrom as such, so long did I not discern the meaning of being enlightened . . . But, monks, when I fully comprehended, as it really is, the satisfaction in the world as such, the misery in the world as such, the escape therefrom as such, then did I discern the meaning of being enlightened in the world. Then did knowledge and insight arise in me . . .'

Enlightenment was the recognition of the world—as it really is: and as it can be, with the use of a little more understanding, through being 'vigilant, strenuous, and resolute.'

VIII

TURNING THE WHEEL OF DOCTRINE

> To set going the Wheel of Doctrine
> To Kasi City now I go;
> And in the blinded world the Drum
> Of the Immortal will I beat.
>
> *Majjhima-Nikaya*

GAUTAMA did not start 'turning the Wheel of Doctrine' immediately after his enlightenment. On the contrary, for a while he hesitated to take this step. According to some accounts he spent about a month in the neighbourhood of the Bodhi-tree, meditating on 'the Chain of Causation in direct and reverse order.' But the probability is that his meditations were not quite so abstract. It is conceivable that he was anxious to be quite clear in his own mind as to the precise implication of certain conclusions. However, he was even more anxious to face the practical issue involved. He was by no means convinced of the wisdom of returning to 'the blinded world to beat the Drum of the Immortal.' Indeed, he entertained serious doubts on the matter; and it took him considerable time and thought to decide what to do next.

His misgivings and hesitancy before undertaking his life-long mission are unequivocally set forth in the formal statements of the *Majjhima-Nikaya*. 'Then I thought,' he says, 'now I have gained the doctrine, profound, hard to perceive, hard to know, tranquil, transcendent, beyond the sphere of reasoning, subtle, to be known by the wise. Mankind is intent on its attachments, and takes delight and pleasure in them. For mankind intent on its attachments . . . it is hard to see the principle of Causality, origination by way of cause. Hard to see is the principle of the cessation of all compound things, the renunciation of clinging to rebirth, the extinction of all craving, absence of passion, cessation, Nirvana. But if I were to teach the Doctrine, and others did not understand it, it would be a weariness to me, a vexation . . .'

These are not Gautama's own words, but they probably represent a fairly faithful picture of the state of his mind after the enlightenment. He was not certain of his capacity for converting people, and, in a way, this uncertainty was characteristic of him. Moreover, in this case, he had a very good reason for his hesitation. Considered from any angle—whether moral, religious, or social—he had arrived at a position, which, though it may not appear startling to-day, was certainly revolutionary in the sixth century B.C. On the moral question, in striking contrast to the arbitrary and empty taboos of conventional moralists, the ethics that he had evolved had their basis in practical psychology. They derived sanction from the concept of certain limits of human experience. He knew that this would give rise to all sorts of misapprehensions. In religious matters, his gospel seemed to deprive the organized hierarchical institutions of their very *raison d'être* by taking no notice whatever of the priestly paraphernalia of elaborate rituals, ceremonies, sacrifices, and charms. He knew that by denying the priestly mediation, as a condition for salvation, he was asking for trouble. On the social side, he was not willing to recognize any distinctions of class and colour, and he was firmly convinced that 'both bad and good qualities, blamed and praised respectively by the wise, are distributed among each of the four classes.' He was aware that if he were to preach an equalitarian doctrine of this nature, it would mean stirring up the hornet's nest of vested interests. What is more, he felt that the fact that he attached little importance to metaphysics and mysticism would render him extremely unpopular with representatives of various metaphysical and mystical schools. On the other hand, although he could easily have enlisted the wholehearted support of the Lokayatikas, or the Worldly-Wise, he had no keen desire to get involved with that boisterously hilarious tribe. 'The springy step' which usually goes with all uncritical flamboyance and enthusiasm, did not particularly appeal to his rather sober bent of mind. The choice, he knew, did not lie between idealistic pretensions and the self-complacency implicit in aggressive and uncomprehending materialism. Somewhere there was a middle way which avoided both; and, as far as he himself was concerned, he was determined not to compromise himself with either of these two extremes. He wanted to place first things first, and was conscious that such a sane and balanced attitude was unlikely to win the approval of his countrymen.

It is not difficult to imagine the kind of thoughts which must have passed through Gautama's mind while he was deciding whether he should go out into the world to preach his doctrine. Naturally he was bewildered by the vast gulf which seemed intellectually to separate him from his contemporaries. The mission which his own destiny had now proposed to him appeared of an overwhelming magnitude and full of risks. He was not sure he would be equal to the task it would impose. He was alone. He felt helpless and weak in face of the exacting demands of his own vision. Finally, however, he decided to take the risk. The Canon, of course, brings Brahma Sahampati down from Heaven to induce Gautama to undertake turning the Wheel of Doctrine so that 'those beings of little impurity that are falling away through not hearing the Doctrine' might be saved. But it would be altogether more appropriate if we gave the whole credit for this decision to the Tathagata's own private judgement.

Gautama had now made up his mind on the crucial question which had been baffling him. He would devote himself to preaching what he considered to be the right way of life. He knew that the task involved risks, though he felt they would be worth while. But before he embarked on this life-long mission he wanted to discuss things with his former teachers, Alara Kalama and Uddaka Ramaputta. Both were men of ripe wisdom, rich in experience and years, well-versed in the art of preaching. He felt it would be useful to have their advice. Also, perhaps, he entertained some faint hopes of converting them to his own cause. True, they were both interested in metaphysics, but the systems to which they adhered were, in their own way, quite novel, advanced, and distinctly revolutionary—at least by the metaphysical standards of the day. There was just a chance that his earnestness might convince them, that whatever the attractions of abiding in the state of Nothingness, or the state of Neither-consciousness-nor-nonconsciousness, it was far nobler to strive for a more humane and reasonable order in the here and now. In any case, he knew, they could hardly grudge him practical advice as to what to do. He had had no word with them since he left them, nor heard of them in the intervening years. His apprenticeship under them had been brief and had ended abruptly; but in both cases the parting had been quite friendly. So he tried to get in touch with them again. Unfortunately, however, they had both been dead some time.

The Canon says that the knowledge of their death was revealed to him in a flash of vision. But we are justified in suspecting that there was also some earthly source of information. Anyhow, whatever may have been the nature of the source from which the information came, it made him change his plans and decide to go straight to Benares. He knew that the five mendicants who had been his admiring companions when he was practising austerities, and who had left him in anger because he had given up the penance and taken to 'solid food,' were now in that city. The *Majjhima* suggests that this was his main reason for going there. The idea occurred to him, it says: 'The five monks did much for me, who attended me when I was intent on striving. What if I first teach the doctrine to the five monks?' This idea might have been at the back of Gautama's mind; but doubtless there were other, far weightier reasons which made him decide in favour of Benares. Since he wanted to preach—and in those days preaching was the only possible way of spreading ideas—there was no better place than Benares in which to begin.

At that time Benares had already become the greatest city of India—a position which it was long to retain. When Megasthenes visited India, it was supposed to have a circumference of eighty-six miles—though probably these measurements are exaggerated. Even in Gautama's days it was by far the largest city within the confines of the Hindu world. It was a flourishing metropolis, 'opulent, prosperous, populous, abundant with food, crowded with people,' a famous seat of Brahmanical learning, a manufacturing centre of unrivalled importance. It was famous for its gorgeous temples and textile fabrics worked with gold, its shrines and metallurgical factories turning out brassware of all descriptions, its rich priests and workers in ivory. Trade-routes converged upon it from every direction. Pilgrims journeyed thither from far-off places, anxious to wash away their sins in the celestial stream flowing from the locks of Siva, the Destroyer and Purifier of the Universe. There were many rich Hindus who looked forward to living there in old age, that they might spend their declining years in an atmosphere congenial to meditation and prayer, and that their funeral rites might be performed on the very banks of the holy Ganges. The vast multitudes who could not afford to die in the City of Felicity tried at least to arrange to have their ashes sent to Benares. The prosperity of its citizens owed not a little to the active trade in the holy water, which, of course, was sent to all parts of India in sealed urns. Yet, though a place of pilgrimage, Benares was by no means dull. It attracted not only penitents, but pleasure-seekers. If it was renowned for its learned exponents of Sacred Lore, it was no less famous for its courtezans, skilled in the art of initiating the pilgrims of pleasure into obscure rites. People came to Benares not only to repent, but to indulge. Indeed, the proximity of the Celestial Stream tended to make both visitors and inhabitants somewhat lax in their morals. Such easy access to Absolution made people bold. After bathing in the 'fountain of sin,' men went for a dip in the Sacred Stream and were pure again. It was all very simple. There was an equal demand for the services of priests and prostitutes. Benares was a gay and lively city.

Thus there were many practical considerations which induced Gautama to journey to the City of Felicity 'to beat the Drum of the Immortal.' The journey itself was quite uneventful, except that Gautama had an amusing encounter with Upaka, an Ajivika ascetic. Upaka, who met Gautama on the road near Gaya, asked him: 'Your faculties, friend, are clear, the colour of your skin is pure and clean. Whom do you follow, friend, in leaving the world? Who is your teacher, and whose doctrine do you approve?'

This was quite a natural question from one mendicant to another. Gautama, on his part, was willing to explain to Upaka what his purpose and intentions were. Probably he told him that he had discovered what lay at the root of human suffering, and a way to the attainment of happiness in the here and now. It is hardly likely that Gautama's answer to Upaka was given in the formal verses which the

Majjhima attributes to him; but a claim of this sort, even when couched in the most modest terms, is apt to sound a little presumptuous—especially to an experienced ascetic.

Upaka was not a credulous child who could easily be taken in. He had been an ascetic long enough to have learnt to temper with cynicism his salvationist ardour. He had doubtless heard many upstart mendicants make similar claims. He had every reason to think that Gautama was yet another lunatic or charlatan of the same species. Suffering or no suffering, he certainly did not care to be duped by a quack or a madman. For himself he entertained no illusions about the possibilities of life, happiness, or 'the extirpation of grief and tribulation in the here and now.' He had long since realized the 'wearisome condition of humanity,' and found a safe anchor in Makkhali Gosala's radical fatalism. He was now quite resigned to his hopelessness, and had no strong wish to try a new 'way,' no matter how rosy its prospects seemed. Death was at least something certain; and he was not prepared to abandon that last certitude of life. He shook his head sceptically at Gautama. As far as Gautama's 'way' was concerned, he was not impressed, and he had little to say. 'Would that it might be so, friend,' was his wise comment. 'Venerable Gautama,' he said, 'your way lies yonder.' He himself significantly turned down a bypath in the opposite direction. Gautama's first serious attempt at conversion had been a failure. But one who wishes to found a 'Kingdom of Happiness' has to take such discouragements philosophically.

Gautama reached Benares 'by gradual journeying.' Whether by accident or design, the first men he was to meet on reaching the City of Felicity were the five monks who had left him in disgust when he had wisely given up his resolve to starve himself to death in search of the Immortal. They were staying outside the city in a park known as the Deer-park of

Isipatana. They had not yet forgiven him his refusal to be a martyr for their sake. As they saw him coming from afar, they said among themselves: 'This, friends, is the ascetic Gautama coming, who lives in abundance, who has given up exertion, and has turned to a life of abundance. We must not greet him, nor rise in respect of him, nor take his bowl and robe, but we will set a seat for him.' However, as he approached and greeted them, they were unable to carry out their decision. They seem to have been moved to remorse. One of them took his bowl and robe, one prepared a seat, another set water for his feet. After which Gautama told them that at last he had gained the doctrine he had sought; that this doctrine, if properly cultivated, would lead to the realization and fulfilment of human purpose 'even in this life.' But the five monks would not hear of it. They were as sceptical of his claims as Upaka. They were ready with their taunts about his living in abundance. They said: 'By that exercise, friend Gautama, by that course and practice of self-mortification, you have not gained that superhuman, truly noble knowledge and insight. Will you, when you now live in abundance, have given up exertion, and have turned to a life of abundance, gain that supernatural, truly noble knowledge and insight?' For them the only possible access to the Immortal lay through a torture-chamber, though for themselves they had the sense to be content with a vicarious realization of 'supernatural, truly noble knowledge and insight.' Gautama tried very patiently to convince them that he had not given up exertion, and that he had not turned to a life of abundance. It was a long and tiresome argument. In the end, however, he persuaded them to accept his good faith and listen to what he had to say. Even then it is doubtful if they were convinced of the underlying truth of his instructions. It seems much more likely that they were impressed by a new note of earnestness and determination in

Gautama's manner; and this must have led them to reason thus: 'This man seems fairly resolute and intent on his mission. He has plenty of zeal and enthusiasm. Let us follow him and see how he carries it out. If he succeeds, we shall have the honour of being known as his first disciples. If he fails—as he probably will—we can easily disown him, as we did before. Our reputation will be none the worse for that. And there is always the chance that he might succeed.'

According to the Vinaya, at this juncture Gautama preached two sermons to the five monks. The account in the *Majjhima* has, however, no place for these pronouncements. It seems certain that, like the Sermon on the Mount, most of Gautama's sermons were never preached in the form in which we find them to-day. As they now stand, they are doubtless synthetic products compiled by Canonical writers and commentators, who wanted to present the basic tenets of the faith in a tabloid and formal shape for the greater ease and comfort of the laity as well as clergy. At the same time, it stands to reason that, having brought them round to his point of view, Gautama enlightened the monks as to the doctrine which he had worked out so carefully; and that in both the discourses there is, in fact, a nucleus which faithfully represents his beliefs on certain fundamental issues.

In the first sermon Gautama begins by exhorting the monks to avoid the two extremes: 'that conjoined with the passions, low, vulgar, common, ignoble, and useless, and that conjoined with self-torture, painful, ignoble, and useless.' He then passes on to the problem of conduct, and defines the Middle Way as an Eightfold Path—the eight factors being 'right views, right intention, right speech, right action, right livelihood, right effort, right mindfulness, right concentration.' Finally, he discusses the broader psychological question of the essential nature of human experience, and enunciates his four basic truths: 'the truth of pain, the truth of the cause of pain, the truth of the cessation of pain, and the truth of the way that leads to the cessation of pain.'

The second sermon concerns itself with a metaphysical issue which has troubled philosophers for nearly three thousand years—the question of the permanence, or otherwise, of the soul. Gautama takes up a position which, for all practical purposes, is diametrically opposed to the Vedantic metaphysicians who held the soul (*atman*) to be a reality behind all psychical phenomena. He dispassionately analyses the various elements of experience, both physical and mental, and observes that, neither individually nor taken as aggregates, are they permanent. By this *reductio ad absurdum* he goes on to emphasize that emancipation from the illusion of permanence is the primary condition for enlightenment.

Taken together, the two sermons bring the psychological character of Gautama's doctrine into sharp relief.

IX

GROWTH OF THE ORDER

GAUTAMA had exaggerated his difficulties. His fear that he would not find people willing to listen to 'the Drum of the Immortal' proved entirely groundless. For the first few months after the foundation of the Order, at any rate, everybody appears to have been enchanted by his preaching. There was no repetition of his unhappy encounter with the cynical Upaka. He found many men and women who were prepared to hail him as the new Prophet. The doctrine which he knew was 'hard to perceive, hard to know, tranquil, transcendent, beyond the sphere of reasoning, subtle, to be known by the wise'—the doctrine which, like the salt-sea, had throughout a uniform taste, the taste of Nirvana; the doctrine which he was afraid would not have the slightest appeal for a humanity feverishly intent 'on its attachments'—this doctrine, strangely enough, caught up the popular imagination.

Miracles are rare—but not impossible. The conversion of the five recalcitrant monks was admittedly a rather lukewarm affair. However, several other conversions which followed were full of fervour and spiritual zeal. There was the case of a youth called Yasa. He was the son and heir of a flourishing and worldly-wise guildmaster of Benares. His youthful years, like those of most rich young men of the day, had been devoted to the pursuit of the Sublime through the avenue of the senses. In a moment of recoil from his life of indulgence and luxury, he decided to take refuge in the new gospel of Renunciation. His father, hearing of his impetuous and foolish resolve, hastened to the Deer-park of Isipatana in order to rescue him from the evil company into which he had fallen. But it was too late. Not only did the guildmaster fail in rescuing his son, but as sometimes happens with those who go to scoff, he himself remained to pray. He found Gautama's instructions so profoundly moving that he begged to be accepted as a lay disciple. Yasa's mother and wife were the next to succumb to the lure of the new Faith. They were the first two women to take refuge in the Buddha. The Wheel of the Excellent Law was now in full swing. More and more citizens of Benares were drawn into the Tathagata's circle. Before three months after his arrival in the Deer-park of Isipatana, Gautama had sixty full-fledged monks in his Order, apart from a large number of lay followers. These sixty initiates were sent out 'to beat the Drum of the Immortal' in the neighbouring countryside. Their mission was prodigiously successful. Soon, so many people were coming forward to take refuge in Gautama's gospel that he was unable personally to attend to all of them. A formula of ordination had to be devised and the monks themselves were allowed to admit newcomers into the Brotherhood.

The rains had now set in, which meant a suspension of missionary activities for three months. Gautama spent this time in comparative seclusion and peace. But as soon as the monsoons were over, work was resumed with renewed vigour. Gautama now decided to move his headquarters to Uruvela. The journey to Uruvela is said to have been eventful; for it was while on the road that, in a spectacular manner, he won over thirty rich young men, who in some later accounts figure as 'the friends of the series of wealthy ones.' These bright young men had gone to the woods to indulge in pastoral love-making. They had their wives with them; and one of them who was still a bachelor had thoughtfully taken a courtezan with him. The courtezan was to cause them some embarrassment. Indeed, she was to be responsible for landing them in the arms of the Prophet of Renunciation. For while these would-be 'friends of the series of wealthy ones' were amusing themselves with amorous frolics in a pleasant grove, the courtezan had taken her chance to steal their belongings. When they discovered what had happened they were naturally much distressed, and immediately set out in search of the runaway. On the road they encountered Gautama and his party. They asked him had he seen the woman. But Gautama, instead of giving them any information about the courtezan, remarked with a familiar irrelevance: 'What do you think, young men: which is better, for you to go in search of a woman, or to go in search of yourselves?' Whether it was that the thirty young men really received a sudden flash of enlightenment, or merely felt a sense of shame at being found in an embarrassing situation, Gautama's thrust is said to have gone home. They readily agreed that it would be far better to go in search of themselves. Gautama then pronounced a discourse and the men were formally admitted into the Order.

It was an infectious doctrine. And it caught laymen and monks alike. After the conversion of the thirty young 'friends of the series of wealthy ones' came the much more important conversion of three brothers named Kassapa. All three were well-known hermits who believed in some occult and magical fire-cult. One lived in the woods of Uruvela, and was consequently known as Uruvela Kassapa; another on the river nearby, and was called Nadi Kassapa, or Kassapa of the river; the third near Gaya, and was called Gaya Kassapa. Presumably, Gautama was anxious to win over the support of these three brothers since between them they had nearly a thousand followers, and no doubt he expected these disciples to follow in the footsteps of their teachers. He succeeded in realizing his purpose, though not without prolonged and tedious argument.

It was to the Kassapas that he preached probably the richest in lyrical spirit of all his discourses. Entitled 'the Sermon on the Lessons to be drawn from Burning,' it is much more familiarly known as 'the Sermon on the Conflagration of Senses,' or simply as 'the Fire Sermon.' It is, in fact, more than a sermon: it is a magnificent poem, which, for its formal arabesque beauty of reiteration, deserves to take its place alongside *Ecclesiastes* and the *Song of Songs*. There is music in it; a music lucid and precise in its meaning, ecstatic in its rhythmic repetition—like a Bach fugue.

We possess a vivid account of the circumstances under which this Sermon was delivered. Gautama was staying at the time in a hermitage on the Elephant Rock near Gaya. Beyond lay the verdant and delightful valley of Rajagaha with its encircling chain of emerald hills. One evening Gautama and his new disciples were sitting on a promontory; they saw a vast flame rise from the hill opposite. The forest was on fire. The conflagration spread with a lightning rapidity till the whole earth, the horizon, and the sky itself seemed to be devoured by flames. It was an awe-inspiring spectacle; everybody was visibly moved. Gautama, who had a habit of building parables of a universal application from things which to others appeared to have no more than an incidental and transitory significance, recognized in the flame an image embodying a deep symbolic truth. He took up the theme and developed it in a slow monotonous tone. He wanted his companions to realize that it was not only the distant forest that was ablaze. The fire in truth, was nearer to them than they could realize. There were nearer things which also were being consumed, but by subtle and invisible flames. The human heart, for instance, was perpetually burning. And not the human heart alone. 'For,' said Gautama to his wondering audience, 'Everything, O monks, is in flames. And what everything is in flames? The eye is in flames. The visible is in flames; the knowledge of the visible is in flames; the feeling which arises from contact with the visible, be it pleasure, be it pain, be it neither pleasure nor pain, this also is in flames. By what fire is it kindled? By the fire of desire, by the fire of hate, by the fire of fascination, it is kindled; by birth, old age, death, pain, lamentation, sorrow, grief, distress, it is kindled . . . This I say . . . Knowing this, O monks, one who is wise becomes weary of the eye, he becomes weary of the visible, he becomes weary of the knowledge of the visible, he becomes weary of the feeling which arises from contact with the visible, be it pleasure, be it pain, be it neither pleasure nor pain . . . This I say . . .'

And so, at once vehement and meditative, he spoke on far into the night, like a person drifting with the impulse of an inexorable mood or thought. And those among his fellow 'wayfarers,' who could visualize the processes behind and beyond the veil of appearances, saw not only the elements of human experience, but the infinite universe of sense and succession in all its potentialities, converging into the likeness of a restless and hungry flame endlessly feeding on itself—placed in a void without beginning, and without end.

The Wheel continued to turn. In the middle of winter, with his now quite numerous contingent of disciples, Gautama decided to move to Rajagaha, capital of Magadha. Apart from missionary reasons, there were a number of practical reasons for this move. The winter in North India is quite severe. No doubt Gautama felt that it would be much more convenient to spend the winter months in some town where certain indispensable amenities of life could be had more easily. The solitudes of Uruvela and Gaya were pleasant to live in. Here it was possible for the Tathagata to abide and meditate in comparative peace. There were no importunate but important householders to pester him with their tales of woe. But for everything and every place under the sun there is a season. And the ill-equipped hermitages in the neighbourhood of Gaya and Uruvela were hardly suitable for residence during the winter months.

The nights were now growing colder. The winds that came from the mountains were as bitter pains. There were frequent frosts. Towards 'the dark half of the month' snow was expected. There was hardly a leaf on the trees. The earth was hard and brittle. The monks shivered in their saffron robes. The problem of the early morning bath was beginning to present increasing difficulties. Gautama was a good psychologist. He was aware that the faith which thinks nothing of moving mountains would often give way under a cold bath. He did not wish to put the faithful to unnecessary trials. Also he was no believer in ascetic or Spartan ideals. If he did not consider 'the higher life' to be contingent on 'nakedness, matted hair, dirt, fasting, lying on the ground, sprinkling the body with ashes, or sitting motionless for long periods,' still less did he think it a matter of cold baths. He had himself no inclination, nor did he demand it of his disciples, to attain Nirvana prematurely by being frozen to death. There have been any number of prophets who have shown great solicitude for their disciples' souls. But few among them have ever been concerned for their physical well-being. Gautama was one of those rarer prophets who combined an interest for the welfare of people's souls with an equally live interest in that of their bodies. Almost the first thing to which he attended on his arrival in Rajagaha was to arrange for the provision of necessary stoves so that the monks could have hot baths. And so it came to pass that, although many incidents in the life of the Buddha were automatically relegated to oblivion, his grateful disciples never forgot to hand down the tradition about the stoves. Gratitude could hardly have gone further. It should be added that as a result of Gautama's emphasis on personal hygiene, Buddhist monasteries, unlike their Christian counterparts, have always been equipped with adequate arrangements for baths.

The hot baths were an important consideration. But it wasn't merely to provide hot baths for the Brotherhood that Gautama had come all the way to Rajagaha. There were other things calling for attention. There was 'the Drum of the Immortal.' Rajagaha was a flourishing, go-ahead city. It was fast rising in importance and prosperity, thanks to the wise policy of King Bimbisara, who believed in commerce even more than in conquest, and so encouraged all kinds of trade—including, as has earlier been remarked, the 'horizontal trade.' Gautama expected his mission to evoke a favourable response among the citizens. The response which he actually met, however, far transcended his expectations. No less a person than the king himself came to visit him. Bimbisara was a hearty monarch, and, as is the way of kings, kept a needlessly large harem. But allowing for this pardonable vanity, he was a genuine and intelligent man. He felt immediately drawn to Gautama, and asked to be accepted as a lay disciple. He invited the whole Brotherhood to dinner at the palace next day, when he himself waited on the Buddha. Further, he endowed a park known as the Veluvena, or Bamboo grove, for the use of Gautama and his Order. This was the beginning of a life-long friendship between the two.

The news that the king had accepted the Buddhist doctrine caused a sensation. So august an example was bound to be enthusiastically followed by his humble subjects. Ever-increasing crowds were now pressing forward to take refuge under the sheltering wings of Gautama's doctrine. The Wheel was gathering momentum at a startling rate. Gautama could not have helped feeling somewhat bewildered by the unexpected success of his mission. On the surface it looked nothing short of a miracle that the doctrine 'profound, hard to perceive, hard to know, tranquil, transcendent, beyond the sphere of reasoning, subtle, to be known by the wise' should have so readily been accepted by the popular mind. Yet the miracle is not altogether incomprehensible. The phenomenal success of the Buddha was due not to the merits of his doctrines, but to the things which it seemed to promise. Those who joined the Order as monks or lay followers seldom troubled themselves about the doctrinal implications of Gautama's gospel. What they saw in it was a magical formula leading to the 'extirpation of grief and tribulation in the here and now,' as well as in the hereafter. There were also other attractions. For the rich it was a new and fashionable craze; to the poor it offered real benefits, since within the Brotherhood no distinctions of caste or rank were observed. Moreover, Gautama enjoined no harsh discipline, no course of self-torture, as a condition for the attainment of the Immortal. Nor, unlike the organized Brahmanical religion, did he demand costly sacrifices and offerings as his commission for acting as go-between in arranging salvation for mortals. Thus the doctrine was not without certain obvious advantages. To laymen it seemed to offer all the beatitudes of a religious life without imposing the obligation of tedious formalities and observances. It was, therefore, natural that at first men should have thronged to take refuge in the new teaching.

Doctrinal subtleties are never the decisive factor in conversions. The quick conversions effected by Gautama were no exception to this law. How much appreciation of the doctrine entered into them may be judged from the story of the conversion of Sariputta and Moggallana. These two men, who were to become Gautama's chief disciples in later years, were serving apprenticeship under Sanjaya, an ascetic living in Rajagaha, who himself claimed some two hundred and fifty odd disciples. Early in life they had pledged a mutual bond that whoever first should catch the Absolute was to tell the other. One day, it is said, Sariputta saw Assaji, an elder of Gautama's Order, going through the street early in the morning on the usual alms-begging round. The elder in question walked with a measured gait. The movements of his arms conformed to the strictest rules of decorum. His eyes were fixed on the ground with becoming modesty. Sariputta was impressed. He was more than impressed. For some inexplicable reason he felt that Assaji was one who had attained the Immortal. Such perfect deportment signified nothing less than 'arahatship.' He followed the elder at a respectable distance, and choosing an opportune moment saluted him, saying: 'Your faculties, friend, are clear, the colour of your skin is pure and clean, whom do you follow, friend, in leaving the world?' Assaji told him that he was a follower of Gautama. Sariputta's next question was naturally about the nature of Gautama's teaching. Here, however, the elder was unable to give Sariputta much information. Decorous behaviour is one thing; the understanding of doctrine quite another. And Assaji, for all his perfect deportment and clear faculties, had never bothered to go into the doctrine which he was supposed to practise. He was blissfully vague as to his master's teaching. Sariputta's awkward question caused him considerable embarrassment. He tried to excuse himself by saying that he was only a novice, and as such it would be a presumption on his part to expound the intricacies of arahatship. Yet far from being put off by Assaji's excuses, Sariputta was charmed. He found the elder's reluctance extremely touching, and interpreted it as a part of the becoming modesty of an arahat. Also, at the back of his mind, there was probably the feeling that Assaji's evasive answer was due to his desire not to share the knowledge of the Absolute. For himself he had no doubt whatever that Assaji possessed that knowledge. He had seen and believed. It was impossible for a man to walk through the streets with such admirable calm without having the Immortal up his sleeve. Therefore he persisted in his inquiry. He said he understood perfectly the elder's difficulties. For himself, he added, he did not want exhaustive information. He would be content with a brief statement of the doctrine. 'Well, friend, tell little or much, but tell me just the meaning—just the meaning is what I want; why speak many words?' he pleaded earnestly. He did not plead in vain. Assaji felt helpless faced with such earnestness. In his confusion, he murmured something about the Tathagata having told 'the cause of all things that proceed from a cause—and also their cessation.' Assaji's statement would not have convinced a child; but it convinced Sariputta. One step was enough for him; and we are told that 'the spotless eye of the doctrine' arose instantaneously in Sariputta.

True to his promise, Sariputta hastened to tell his friend Moggallana of the happy event. In credulity Moggallana had little to learn from Sariputta; and it did not take long for 'the spotless eye of the doctrine' to arise in him too. And not in him only. The other disciples had only to be told to be persuaded into accepting the new gospel. The spotless eyes of the doctrine were springing up everywhere in alarming fashion. The Wheel was turning.

But it is in the very nature of enthusiasm that it cannot last. Signs that the popular effervescence was dying began to appear even during the two months that Gautama stayed at Rajagaha. There were no spectacular mass-conversions following the conversion of Sariputta, Moggallana, and Sanjaya's two hundred and fifty pupils. Indeed, there had been some trouble over that affair. When his faithless disciples had told him of their intention to transfer their allegiance to Gautama, and had given him friendly advice that he should likewise become a devotee of the Buddha, Sanjaya had been furious. Fortunately the language in which he expressed his opinion of the new Messiah, who had enticed all his pupils away from him, has not been recorded. But it is recorded that hot blood came from his mouth when he saw his disciples leaving him.

Sanjaya was not the man to forget an injury. He took the desertion of his pupils very much to heart, and lost no opportunity of discrediting Gautama and his Order in every possible way. He spoke bitter verses attacking the Tathagata and his monks. He broadcast these throughout the city by word of mouth. Soon they were on the lips of every street urchin. The purport of these verses was that 'the ascetic Gautama is come to bring childlessness; the ascetic Gautama is come to bring subversion of families. Already hath he turned ten thousand hermits into his disciples, and he hath enticed two hundred and fifty mendicant-disciples of Sanjaya into his own Order. And now many distinguished and noble youths of the Magadha kingdom are betaking themselves to the ascetic Gautama to lead a religious life. The ascetic Gautama is intent on depopulating and ruining the country.'

Sanjaya's propaganda achieved its end. An atmosphere of tension was created in which it was impossible even for the Tathagata to escape ridicule and abuse. Though he continued to enjoy the confidence and friendship of the king of Magadha, he lost some of his popularity with the general public. His men began to be coldly received, if not with open hostility. Worldly-wise householders avoided the Buddhist missionaries. When the joys of 'arahatship' were painted to them in glowing colours, they calmly confessed their preference for the simpler pleasures of conjugality and parenthood. When members of Gautama's Order went to beg for alms, they found angry wives shutting doors in their faces. They were not going to feed men who designed to entice away their husbands. Apprehensive hermit-philosophers and mystagogues wondered if their own disciples would next desert to join Gautama's crusade. Sanjaya was getting his own back.

These angry demonstrations were ominous. They marked the beginning of the long polemics between orthodox Brahmanism and Buddhism destined to last for several centuries, and indeed until the virtual disappearance of Buddhism from India. The organized dispensers of religious benefits—the Brahmans—had begun to voice their disapproval. Ruling hierarchies—whether of temporal or spiritual order—have at all times considered reformers a nuisance. The Brahmanical Dispensation of that day was no exception to this rule. At first the priestly order had not paid much attention to Gautama's missionary activities. They had been inclined to regard him as yet another of those whimsical but harmless mystics with whom they were fairly familiar; no doubt they had hoped that eventually he would settle down with a few kindred spirits in some forest-hermitage. This they would not have minded at all. They were powerful enough to afford a few heretics who had, moreover, the saving grace of being amusing. But they soon discovered that they had miscalculated. Gautama's challenge went beyond the limits they could safely tolerate. Success for his crusade, they were quick to realize, would mean complete ruin for them and their kind. And already their position was far from secure. There were the sceptics who were constantly causing them embarrassment by demanding that they should produce Brahma in flesh and blood. There were the Lokayatikas who libellously described their ritual and ceremonies as frauds. They could not afford yet another radical, even if indirect, attack on their authority.

They found Sanjaya a godsend. They took up his refrain with alacrity, and accused Gautama of being responsible for mass-desertions by husbands of their wives—and, consequently, being instrumental in causing childlessness. With an unfailing shrewdness, characteristic of the Brahmanical genius, they saw at once that this was an issue on which they could make common cause with the whole patriarchal system. At heart no true patriarch had anything but contempt for the effeminate monastic ideal of celibacy. The Brahmans, on their part, did not reject celibacy outright. Indeed, they gave it a definite place in their scheme of things. But they insisted—and the qualification was of paramount importance—that for everything there is a season, a time for every purpose under heaven. There is a time to procreate, and a time to be a celibate. And youth and middle age are not the times for celibacy, from a Brahmanical standpoint. By common consent these are the years best suited for procreation. Old age is the proper time and season for undertaking such vows, it was sensibly suggested. When one has done one's duty towards society and produced enough offspring to ensure the continuation of the race, one is fully entitled to do penance, which leads to higher beatitudes. But not before.

Such were the sane views held by the Brahmanical hierarchy. And they were not held on arbitrary grounds. The warrant for these opinions was derived straight from Brahma's mouth. Biblical law-givers were not original and unique in commanding the children of Israel to beget and multiply. The Hindu law-givers had forestalled them in this practical wisdom. Manu's Sacred Laws had left no ambiguity on this point. And the Brahmans never forgot to lay special emphasis on the necessity for the systematic, continuous, and resolute prolification of the species. 'Go ye and beget and multiply,' they exhorted the laity. They did more than merely exhort. Example is the best precept, and the Brahmans themselves set the pace. It was eminently desirable that men should beget and multiply. The intricate, almost endless ceremonies connected with these natural processes, formed the chief source of priestly revenue. 'The Order of Philosophers' was interested in many other things besides philosophy. For instance, it was vitally interested in births, copulations, and deaths. Most of the solemn sacrificial ritual was intimately linked with these cardinal events of human life. A birth meant free meals and various other rich gifts for the priests. When they consecrated a marriage, gold and silver poured into their sacerdotal pockets. At death they arranged for the well-being of the departed soul in the next world—but not without charging adequate fees. It did not need much philosophy to see that a falling birth-rate would inevitably be reflected before long by a corresponding decline in the number of marriages and deaths. It was the simplest arithmetic. Gautama had himself admitted that this best of all possible worlds is a world of 'dependent origination.' As such there was no choice. To keep the Wheel of Becoming in motion, every one must beget and multiply. It was essential for the livelihood of priests. For what would become of them were a large number of householders to forsake their homes and take to a monastic life?

It was a matter of the utmost gravity. The finances of the Brahmanical Dispensation were at stake. The very foundation of the patriarchal society was threatened. The astute leaders of the priestly and patriarchal interests were in perfect accord on this question. This craze for celibacy must be nipped in the bud. The contagious doctrine must at all costs be prevented from spreading. It was all very well for

Gautama's newly-ordained 'arahats' to rhapsodize about the delights of 'Disburdenment.' The priests and patriarchs also desired 'Disburdenment,' though naturally of a different kind. Everybody has his special problems. The priests and patriarchs had theirs. Most of them had their nubile daughters. This was a heavy responsibility. How could they ever find 'worthy wooers' for the 'ripe unwedded maidens' if all the eligible young men of Magadha were to be lured away into the Buddhist Brotherhood by the charms of 'arahatship'? It was a question well worth some serious reflection. No householder wished to keep a virgin daughter in his house after she had attained puberty. Gods above looked very unfavourably on any attempt to prolong virginity. The Sacred Texts said so explicitly; they made it imperative for parents to find suitable husbands for their daughters when they had reached the crucial age of ten. Failure to conform to this rule entailed very disagreeable consequences, not only for the offending parents, but for all their remote ancestors in heaven. Parolles in *All's Well That Ends Well* remarks: 'It is not politic in the commonwealth of nature to preserve virginity.' This is a belief the Hindu Law-Givers have echoed all through the ages. The ancients, like Vasistha and Yajnavalkya, were still more vehement in their condemnation of the virgin state. They had no use for it whatever. They regarded it as something contrary to the Divine purpose, and maintained that, preserved beyond a certain age, it was not only impolitic, but definitely sinful.

The priests and patriarchs of Rajagaha had reasons to feel alarmed. They had nothing against the Tathagata personally. Those among them who had met him, held him in high esteem. He was gentle and compassionate; he had charming manners; as a man they had nothing but praise for him. When they met him, orthodox Brahmans themselves could not help feeling a measure of admiration. The more advanced and catholic thinkers in the Brahmanical hierarchy would probably have confessed, in private at any rate, that there was a great deal of truth and wisdom in what he said. But truth and wisdom are not always expedient. Indeed, expediency often demands the sacrifice of truth and wisdom. Gautama's teaching raised certain practical issues bearing on the general good. Here the priests and patriarchs differed vitally from him. They had their own ideas for the general good. The thorny question of celibacy was a typical instance.

The Tathagata was by no means blind to the practical issue. The opposition to his teaching was the result of misunderstanding. He had never conceived the Middle Way as being fundamentally incompatible with propagation. He did not want to make every one a monk. To avoid misunderstanding he had been careful not to include celibacy among the conditions of his noble Eightfold Path. Further, from the very beginning, he had drawn a clear line of distinction between a discipline to be practised by the many, and a discipline to be followed by the chosen few only. These latter, it is true, were to consider themselves as dedicated men, and were expected to renounce all personal ties. But there was nothing unreasonable or perverse in this demand. To serve a great cause it is often necessary to abandon all longings for personal fulfilment. The cause to which Gautama desired to devote himself and his Order was certainly worthy of such sacrifice. However, he did not expect—and did not even desire—this sacrifice from the multitude. For this reason he had been anxious to emphasize the difficulties of his doctrines, rather than its promises. If, in spite of all these precautions, some householders still seemed to be eager to renounce their attachments, it was not the Tathagata's fault.

Gautama explained all this in his defence; but it was no good protesting. It is difficult to catch up with a lie. In polemics the first shot is often half the battle; Sanjaya and the vested interests who supported him had won the first round. Gautama was unable to counteract the misrepresentations which they had spread. Despite his personal charm and his influential friends, he was subjected to characteristic Brahmanical abuse, 'copious, not at all stinted.'

Within two months Gautama had come to the conclusion that a temporary retreat from Rajagaha might best serve to kill the controversy. An opportunity for such retreat had also presented itself. He had received an invitation from his aged father to visit Kapilavastu. Although Gautama had never communicated with his people since he left home, they had heard of him from time to time through various sources. After his enlightenment, he had suddenly achieved celebrity. The news of his success had doubtless reached his family. It was not their notion of success, but they could no longer afford to ignore him. Indeed, they were anxious to take up the broken threads, even to make some kind of public demonstration of their lasting affection and regard for the prodigal. Among his followers the prodigal now counted such exalted personages as the king of Magadha: it was obviously time to recognize him.

Gautama went to Kapilavastu. He and his mendicants stayed in a grove outside the town. His father and kinsfolk came to see him, but there was no killing of the fatted calf. It seems that they were chagrined by the sight of a shaven monk in yellow robes; it revived painful memories of all the tribal and paternal hopes he had betrayed; and they left without providing a proper meal for the prodigal and his companions. Gautama, however, was not very much perturbed by this impoliteness. Next morning he set out to beg food from house to house, as was his custom. Suddhodana soon came to hear that his son was asking for alms in the streets like any motley beggar. It is said he was deeply moved. But even more than being moved, he was alarmed perhaps. He was probably no longer the President of the Republic, but he was still an eminent and highly-honoured citizen of Kapilavastu. He had his own and his family's reputation to consider. He did not want his son to be seen begging in the streets of his native town. What would the people say? He hastened to find Gautama, and after some persuasion, brought him to his house.

All the members of the household were present, with one conspicuous exception—'the ruddy goose parted from her mate.' Suddhodana explained the reason of his daughter-in-law's absence. He said that she had let it be known: 'If I have any value in his eyes, he will come himself to my presence. I can welcome him better here.' Gautama understood the subtle hint. He went to her chamber accompanied by two of his mendicants whom he previously cautioned that should she try to 'do reverence' to the Tathagata in a manner which might not be strictly in conformity with the rules and regulations of the Order, they should not prevent her from so doing, nor show any bewilderment at her conduct. Fortunately, however, the contingency did not arise. The lady did not make any embarrassing demonstration of her affection for her long-lost husband. She was sensible enough to realize the gulf which divided her from Gautama; and even if she had entertained any ardent designs, the spectacle of a recluse in rough garments, with a shaven head and a grave and weather-beaten face, must have been sufficient to make her abandon them. She respectfully clasped his feet and wept.

The reconciliation between the prodigal and his family

was now complete. Nevertheless, Gautama's family could not fully approve of his mode of living. There is no real basis of understanding between those who are worldly-wise and those for whom the process of 'wayfaring' through the world is itself the end: the two ways are fundamentally incompatible. Suddhodana, though he was very fond of his son, found his behaviour both strange and shocking. Moreover, during his stay in Kapilavastu Gautama did a number of things which were liable to strain his relations with his family still further. It is recorded, for instance, that he persuaded Nanda, his half-brother, and son of Mahapajapati, whose coming-of-age ceremony had just been celebrated, 'to leave the world'; also that he conferred 'Disburdenment' on his own son, Rahula, by admitting him to the Order.

The account of this latter incident is at once pathetic and amusing. After Gautama had been about a week in Kapilavastu, his wife had the happy inspiration of dressing up the child in his best clothes and saying to him: 'See, dear, that golden-coloured ascetic . . . He is your father, and he had four great vases of treasure, but since he left the world we do not see them. Go and ask for your inheritance . . .' Rahula did as he had been instructed. While Gautama was having his meal the boy walked up to 'the golden-coloured ascetic' and said lovingly: 'Pleasant, ascetic, is your shadow.' He then asked for his inheritance. Gautama, however, did not pay any heed to the child's demand and went on eating. Having finished the meal he rose to go to the Nigrodha park where he was staying. Rahula followed him, still asking for the 'four great vases of treasure' which his mother had told him had been hidden. On his part, Gautama maintained an enigmatic silence with regard to the hidden treasure. When he reached the Nigrodha grove he asked Sariputta to 'disburden' the unsuspecting Rahula, and bestow upon him the priceless gift of the doctrine. This was done and Rahula had his inheritance, though it was not the kind of treasure for which either the child or his mother could feel very grateful.

Both these acts of Gautama were hardly calculated to promote understanding between his father and himself. Suddhodana was very much upset by the loss both of his second son and grandson. But he was too old to pick new quarrels with his son, and too heart-broken to make angry scenes. All he could do was bemoan: 'When you abandoned the world, it was no small pain to me, so when Nanda did, and especially so in the case of Rahula. The love of a son cuts through the skin, having cut through the skin it cuts through the hide, the flesh, the sinew, the bone, the marrow . . .' He begged Gautama not to confer ordination 'on a son without the permission of his mother and father.' This request Gautama readily granted: he had no intention of enticing sons from their parents. Rahula, being his own son, was a special case.

The prodigal's visit to his parental home came to an end soon after these unfortunate incidents. On his way to Rajagaha he stayed for a time at Anupiya, on the banks of the river Anoma, in the neighbourhood where he had dismissed his charioteer on the historic night of his Renunciation. It was a fruitful sojourn, and there were several additions to the Brotherhood. The most important among them were Ananda, Devadatta, Anuruddha, and Upali. The first two were Gautama's cousins. The former was to become the Tathagata's most intimate and faithful companion; the latter was to achieve notoriety as his most insidious rival. Anuruddha afterwards distinguished himself by his mastery of the Buddhist metaphysics. And the last, Upali, a barber by birth and profession, was later on to display the same skill in intellectual hair-splitting which he had hitherto shown in hairdressing, shaving, and other kindred crafts.

The situation in Rajagaha had eased considerably. Gautama spent the rainy season—the second after his enlightenment—at Rajagaha, where he found the people once again sympathetic to the doctrine.

X

YEARS OF WANDERING AND A DAY

GAUTAMA'S ministry lasted for nearly half a century. No connected chronicle of this long stretch of time is available, and perhaps none is needed. In the accounts of his life as a wandering preacher, we may agree with Oldenberg, 'unquestionable truth is mixed up with just as much unquestionable romance.' To separate one from the other is in itself a difficult task; to attempt to construct anything like a definite and reliable chronology is attempting the impossible. However, the more credible episodes connected with the first twenty years of his wandering can be seen to form some kind of a vague sequence.

After spending the rainy season at Rajagaha, Gautama visited Savatthi. This visit was highly opportune. For Savatthi was at the time one of the most flourishing cities in India, being the capital of King Pasenadi of Kosala, who shared with King Bimbisara of Magadha the balance of political power in the valley of the Ganges. Gautama's visit to Savatthi was made possible through the mediacy of Sudatta, a rich and highly-honoured householder. Sudatta had come to transact some business in Rajagaha, and to see his sister who was married to a guildmaster in that city. The latter happened to be a lay follower of the Tathagata, and it was through her that Sudatta met the Buddha. Sudatta was quickly converted to the new faith, and before leaving Rajagaha he invited Gautama to his native city, offering to make all arrangements for the journey as well as for his stay there. Gautama, who was above all a practical man, readily accepted the offer. His next Retreat, consequently, was spent at Savatthi. Sudatta left no stone unturned to make this stay a success. He even bought a very beautiful grove known as Jetavana, built a monastery on this pleasant site, and presented it to the Brotherhood. A grateful Buddha acknowledged this magnificent gift by conferring on Sudatta the title of Anathapindika, which means literally 'giver of the alms to the unprotected.' The commentators on 'the Story of the Lineage' even add that, on the occasion of accepting Sudatta's gift, Gautama dwelt at some length on the advantages of monasteries, saying:

Cold they ward off, and heat;
So also beasts of prey,
And creeping things, and gnats,
And rains in the cold season,
And when the dreaded heat and winds
Arise, they ward them off.

To give to monks a dwelling-place,
Wherein in safety and at ease
To think and insight gain,
The Buddha praises most of all.

Let therefore a wise man,
Regarding his own weal,
Have pleasant monasteries built,
And lodge there learned men.

Let him with careful mien,
Give food to them and drink,
And clothes, and dwelling-places
To the upright in mind.

Then they shall preach to him the Norm—
The Norm, dispelling every grief—
Which Norm, when here he learns, he sins
No more, reaching the perfect well.

This discourse—the intentions of which are clear enough—is evidently a happy invention of the commentators themselves. But it is certain that Gautama spent many a Retreat in the monastery at Jetavana, and that Sudatta figures in the list of the eighty chief disciples under the well-merited title of 'chief of almsgivers.'

The fourth year of Gautama's ministry is noteworthy for two events. One, that Gautama was called in to settle an irrigation dispute between the Koliyans (his mother's clan) and the Sakyas. His arbitration succeeded in reconciling the two clans. The other was the conversion of Uggasena, the son of a rich guildmaster of Rajagaha who had abandoned his wordly estate and turned rope-dancer for the love of an acrobat's daughter.

Next year Gautama had to pay a hurried visit to Kapilavastu to see his dying father. After performing the obsequies of Suddhodana, he returned to Mahavana in the neighbourhood of Vesali, the capital of the powerful confederation of the Licchavis. Mahapajapati, his aunt and step-mother, as also his own wife and some other distinguished ladies from Kapilavastu, are said to have followed him to Vesali. So far Gautama had allowed women to become his lay disciples, but he had withheld from them the privilege of complete 'Disburdenment.' Mahapajapati and her companions, however, begged to be allowed 'to leave the world,' to take full refuge in the Tathagata's doctrine and discipline. This naturally would have involved the founding of an Order of female mendicants; and Gautama, for various reasons which are not difficult to understand, was rather reluctant to take such a step. Eventually he granted Mahapajapati's request, but only on certain strict conditions. He is even reported having remarked to Ananda, whose advocacy was principally responsible for the founding of the Order for women: 'If women had not received the going forth in the doctrine and discipline, the religious system would have lasted long, the good doctrine would have stayed for a thousand years; but as women have gone forth, now the religious system will not last long, now, Ananda, the good doctrine will last only five hundred years. Just as houses, where there are many women and few men, are easily broken into by robbers, even so in the doctrine and discipline in which a woman goes forth the religious system will not last long . . .' The cynicism implied in this remark probably voices the sentiments of some latter-day Buddhist anchorite rather than of Gautama himself; for it is completely incongruous with an earlier statement of Gautama apropos the same incident, where he admits to Ananda that a woman is in every way 'capable' of realizing 'the fruit of Entering-the-stream, of the Once-returner, of the Non-returner, and even of the Arahatship.'

No doubt Gautama had cause to hesitate before admitting women into the Order, but his hesitation was not actuated by any contempt for them as such. And although the prophecy contained in the above lament came remarkably near being true—and, indeed, one might say that the dissolution of the doctrine and discipline started even earlier than the prophet had predicted—it is only fair to state that the blame for this unhappy contingency must be apportioned equally among the male and female mendicants.

Gautama spent his sixth Retreat at Makula hill, near Kosambi (modern Allahabad). When the rains were over, he returned to Rajagaha and admitted Bimbisara's wife, Kshema, to the Order. Gautama's popularity with people in high places could not fail to excite jealousy. There were frequent attempts by his enemies to discredit both himself and his creed; two serious attempts of this nature are recorded as having been made during the seventh year of his ministry. On one occasion, when Gautama was staying at

Jetavana monastery, a rival sect induced a woman named Chincha to visit the Buddha, and afterwards to feign pregnancy and accuse the Tathagata of having been the cause of her undoing. They seem to have forgotten that pregnancy is not a thing that can be feigned, and consequently their conspiracy was quickly exposed. Another story tells of a still more serious attempt to bring Gautama into disrepute. Members of a hostile organization persuaded a woman known as Sundari to pay frequent visits to Jetavana. They then had her killed and her body thrown in Gautama's monastery. Fortunately, in a drunken brawl, the hired assassins blurted out their crime and were apprehended by the king's spies. It is difficult to determine how much truth there is in these stories, but on the face of it they seem credible. If they are true, then Gautama's life during his Buddahood could not have been quite so Nirvanic as has usually been represented.

The eighth rainy season was spent at Sumsumaragiri, a town near Kapilavastu. It was here that Gautama converted Nakulapita and his wife. The conversion took place under strange circumstances. The couple had the original idea when they saw Gautama of claiming him as their long-lost son, saying to him: 'Son, you have left us for such a long time, where have you been living?' Gautama, though he must have felt embarrassed by this unexpected recognition, is said to have admitted their claim. But, significantly, he explained that he had been their son not in his present, but in some previous birth. This remark is often produced as a proof of Gautama's belief in reincarnation. But it is impossible to examine it objectively without concluding that the Tathagata was merely respecting the feelings of a credulous but well-meaning couple.

Next year a still more curious adventure befell Gautama. He was staying at Kosambi. Magandiya, an ambitious and opulent Brahman, had a daughter to marry. She had reached the critical age, and the worthy Brahman was anxiously on the look-out for a suitable 'wooer.' He happened to see the Buddha, and noting him to be sound in wind and limb decided that he would make an excellent match. He decided to speak to Gautama. Magandiya's wife, who appears to have been rather a sensible person, tried to dissuade him from making a fool of himself, and even told him, that as far as she knew the Tathagata was not interested in women, and therefore it was futile to offer him their daughter's hand in marriage. But the ambitious Brahman was not to be dissuaded so easily. So he went to Gautama, put his proposal before him with all the vehemence and earnestness of which he was capable; and it was with some difficulty that Gautama at last got rid of this importunate and insistent Brahman. Magandiya's daughter, when she came to know that Gautama had refused her, was gravely chagrined. She took the refusal as a personal slight, and never forgave him. It is even recorded that later, when she was happily married to Udena, the ruling chief of Kosambi, she had her revenge. Discovering that another lady in Udena's harem, Samavati, was a devotee of the Buddha, she conspired against her and brought about her death by having her palace set on fire.

The tenth year of Gautama's ministry was a fateful one in the history of the Order: it saw the first breaking out of serious dissension in the Brotherhood. An unknown monk of Kosambi is said to have been the cause. He had committed some minor transgression, yet when it was pointed out to him, he refused to admit it. Now according to the existing rules a man could not be regarded as a transgressor unless he himself recognized his offence. In spite of this, however, the monk was excommunicated. A heated controversy followed. Some members of the Order maintained that the excommunication was unjustified; others held that not to have taken some drastic action against the impenitent monk would have been tantamount to putting a premium on evil. Eventually the matter was referred to Gautama. He tried hard to restore peace between the two parties. 'Do not look long,' he said, 'do not look short; for not by hatred are hatreds calmed; by non-hatred are hatreds calmed.' But all his attempts towards peace failed. The dispute rose to such heights that Gautama was forced to retire to the solitudes of Parileyyaka forest, leaving the querulous mendicants to fight out the issue amongst themselves. It was three months before the monks came to their senses and went to Savatthi begging Gautama's forgiveness.

There were only two important conversions in the eleventh and twelfth years—those of Bharadvaja and Veranja. The twelfth year was also marked by the outbreak of a serious famine in the Gangetic valley. This could not help having unhappy repercussions for the Brotherhood whose members were dependent upon the charity of the householders. Food was scarce and the World-Renouncers seem to have had a lean time. It is recorded, for instance, that Veranja had invited the Tathagata to spend the Retreat with him, only to find it impossible to keep his promise; he got out of it by conveniently forgetting all about it till the return of better times, when he sought out the Buddha and apologized for his forgetfulness.

The thirteenth and fourteenth years were both uneventful. The former was spent at Calika hill; the latter at Savatthi. Next year Suppabuddha, Gautama's father-in-law, had a fatal accident trying to stop a runaway horse. It is doubtful if Gautama felt this death as a great personal loss. Ever since the day he had left his house and thus condemned his wife to what was virtually a life-long widowhood, relations between himself and his father-in-law had been strained. Gautama, it is true, harboured no ill-feelings against Suppabuddha, but the latter had never forgiven the Tathagata for deserting his daughter. Even time had failed to heal the breach between them; Suppabuddha's bitterness against Gautama had rather become intensified with the years; and only a week before he died Suppabuddha had been drunk and cursed Gautama publicly in the streets of Kapilavastu.

The sixteenth rainy season was spent at Alavi, the seventeenth at Rajagaha, the eighteenth at Calika Hill, and the nineteenth at Rajagaha again. This period was not marked by any important happenings. In the twentieth year of his ministry Gautama converted Angulimala, an infamous robber, the terror of the countryside round Savatthi, who was known to wear 'a garland of his victims' fingers.' When King Pasenadi came to hear of this remarkable conversion he is said to have 'complimented Buddha as the tamer of the untamed.'

It was in the same year that Ananda was appointed Gautama's permanent personal companion. This appointment was made under somewhat curious circumstances. The custom hitherto had been that the mendicants in the Tathagata's entourage attended him each day in turn. The system had worked satisfactorily until one day Gautama was treated contemptuously by the elder Nagasamala. Gautama and the elder were walking down a country road when they came to a crossroad. As though to assert his independence, Nagasamala pointed in one direction and said, 'that is the way, lord, we will go by that.' But as it was not the direction which Gautama wished, he pointed the other way and said: 'No, Nagasamala, this is the way—we will go this way.' Nagasamala was possibly in a churlish mood: he refused to take the Tathagata's gentle hint. In vain Gautama tried to reason with the elder, but his gentleness was wasted, and Nagasamala angrily put down Gautama's bowl and robe. 'Here, lord,' he said, 'is your bowl and robe,' and so saying, departed. Gautama was naturally distressed by this incident; and though at the time he said nothing about it, he explained to the Brotherhood at Savatthi later, that now he was advancing in years, he felt the need of a permanent companion. All the mendicants, including Sariputta and Moggallana, offered their services. Ananda, the last to rise and offer himself, was the one to be accepted. For the rest of the Tathagata's life, the last enjoyed the privilege of being the first.

It was a monotonous journey. Days passed; seasons changed; but there was hardly ever any variation in the normal routine of the Tathagata's life. Years followed each other in a long but unexciting procession. The records of the last twenty-five years of Gautama's ministry are far too vague to form a narrative; all that can be said is that they were spent in wandering from place to place preaching his philosophy of life. It was an age of wandering preachers. The whole country was alive with a spiritual and intellectual restlessness seldom surpassed, and only rarely equalled, in later history. Some idea of the interest in abstract problems shown by the people in general may be formed from the fact that not only men, but women ascetics were in the habit of taking to a wandering mode of life, and we may regard as typical the case of the woman-preacher who 'was in the habit of going from village to village, and setting up at the entrance to the village a broomstick with the announcement, that she was willing to discuss with any one who should overturn it.'

Gautama himself wandered a great deal. Even when very old he spent the best part of the year travelling from place to place. His yearly round was confined to the valley of the Ganges, that is to say, the country of the Magadhas, the Sakyas, the Kosalas, the Licchavis, the Vamsas, and the Vajjians. The rainy season and the mid-winter months were spent in Retreat. These Retreats were kept at either of the monasteries built by Gautama's wealthy patrons at Rajagaha and Savatthi for the use of the Order.

Buddhaghosa has left an account of the Tathagata's daily life, which, when divested of its euphony and supernatural details, gives what is probably a true picture. 'The Blessed One . . .' he writes, 'used to rise up early (about five in the morning) and, out of consideration for his personal attendant, was wont to wash and dress himself without calling for any assistance. Then, till it was time to go on his round for alms, he would retire to a solitary place and meditate. When that time arrived, he would dress himself completely in the three robes (the dress which every one in the Brotherhood was expected to wear), take his bowl in his hand and, sometimes alone, sometimes attended by his followers, would enter the neighbouring village or town . . .' The procedure of going out to beg alms, however, was only formal. For the Tathagata had many well-to-do lay followers, and these appear to have vied with each other in feeding him, as well as his mendicants. Buddhaghosa writes: 'Then clad in their best and brightest, and bringing garlands and nosegays with them, they [the rich lay followers] would come forth into the street and, offering their flowers to the Blessed One, would vie with one another saying: "To-day, sir, take your meal with us; we will make provision for ten, and we for twenty, and we for a hundred of your followers." ' Even a far less compassionate person than the Blessed One would have found it difficult to refuse such warm solicitations, and it is not surprising that Gautama readily accepted the invitation to a meal and let his host 'take his bowl, and spread mats for him.' After the necessary ritual of taking 'some solid food' was over, there followed the inevitable discourse. In his sermons, we are told, the Blessed One paid due regard to the 'spiritual capacity' of his audience. 'And when he had thus had mercy on the multitude,' continues Buddhaghosa, 'he would arise from his seat and depart to the place where he lodged. And when he had come there, he would sit on the open verandah, awaiting the time when the rest of his followers should also have finished their meal. And when his attendant announced they had done so, he would enter his private apartment. Thus was he occupied up to the midday meal.'

Afternoons were spent in exhortation to the brethren in the fold to follow the Eightfold Path. Then some among the flock 'would ask him to suggest a subject for meditation suitable to the spiritual capacity of each, and when he had done so, they would retire each to the solitary place he was wont to frequent, and meditate on the subject set. Then would the Blessed One retire within the private chamber, perfumed with flowers; and calm and self-possessed, he would rest awhile during the heat of the day. Then when his body was reposed he would arise from his couch and for a space consider the circumstances of the people near, that he might do them good. And at the fall of the day, folk from the neighbouring villages or town would gather together at the place where he was lodging, bringing with them offerings of flowers. Seated in the lecture-hall, in a manner suitable to the occasion, and suitable to their beliefs, he would discourse to them of the Truth. Then, seeing that the proper time had come, he would dismiss these folk, who, saluting him, would go away. Thus was he occupied in the afternoon.'

Generally the day ended as peacefully as it began. At sunset 'should the Blessed One feel the need of the refreshment of a bath, he would bathe, while some brother of the Order, attendant on him, would prepare the divan in the chamber, perfumed with flowers. And in the evening he would sit alone, still in all his robes, till the brethren, returned from their meditations, began to assemble. Then some would ask him questions on things that puzzled them, some would speak of their meditations, some would ask for an exposition of the Truth. Thus would the first watch of the night pass, as the Blessed One satisfied the desire of each, and then they would take their leave. And part of the rest of the night he would spend in meditation, walking up and down outside his chamber; and part he would rest lying down, calm and self-possessed, within . . .' And before long another day would dawn, and there would be another turn of the Wheel.

Of course, Buddhaghosa has idealized. In actual practice the routine of the Blessed One's life probably did not work out in quite so idyllic a manner. Buddhaghosa has evidently forgotten the daily embarrassments which must beset the head of a quasi-religious Order, and from which Gautama was by no means immune. He has forgotten, too, the prolonged and painful polemical discussions into which Gautama was often drawn, much against his will. And his account obviously takes no cognizance of the misrepresentations to which the Blessed One was frequently subjected by certain rival sects and schools who wished to discredit the man and his philosophy. To ignore these aspects of Gautama's life is to leave the picture incomplete. And yet Buddhaghosa's picture, partial as it is, yields an impression which somehow seems to correspond truly to the inner curve of Gautama's life. Fundamentally his manner of living was absolutely consistent with the 'way' he had postulated: it was peaceful, quiet, profound. Beyond the difficulties and embarrassments which arose from his intercourse with the world of men, beyond the long-drawn polemics in which he was involved, because of the strange simplicity and newness of his doctrine, and beyond the insufferable fools whose misunderstandings he had to suffer so patiently and for so long, there was in him a quietude, a core of serenity which nothing could disturb. The period of his ministry was no doubt full of incidents; yet it would not be a distortion of facts to suggest that essentially it was an uneventful period. For there is in the life of each individual a decisive moment, a final turning-point which determines one's particular path through this world. After that crucial event, the pattern of one's being admits of no further change. One's awareness, it is true, may become more comprehensive, and one's thought more mature, but there is no qualitative change in the nature of one's awareness and thought; in other words, though one may 'know more, one cannot know *differently*.' In Gautama's psychological development this decisive turning-point came at the time of his enlightenment. The conclusions which he then formed did not undergo any significant change during the rest of his life. The horizon of his vision continued to expand, to become more inclusive and sharply outlined, as the years passed. But it remained, to the end of his days, the same horizon and the same vision.

It was also in some ways a lonely journey. In the very nature of things, it could not have been otherwise. For a person who represents what Professor I. A. Richards describes as 'the most aware point of his age' is inevitably so far in advance of the understanding of his contemporaries, that he cannot help feeling an intimate sense of isolation. A life-long solitude is the penalty of all genius; and Gautama, too, had some experience of this kind of personal solitude. No man could have been kinder, more considerate and gentle, in his dealings with his fellow-men. But his relationships were nearly always kept on an impersonal plane. During the whole of his life he does not seem to have formed any personal attachments. Even Ananda, who for nearly a quarter of a century was his constant companion, could not break through that ultimate barrier which separated him from his Master.

But Gautama was by no means friendless. He commanded not only respect and admiration, but fervent devotion. He had a very wide circle of friends. These were drawn from all quarters, high as well as low. In choosing his friends and acquaintances, Gautama 'ignored completely and absolutely all advantages or disadvantages arising from birth, occupation, and social status.' He had no use for snobbery of any kind, and lost no opportunity of discouraging it. Powerful potentates and lowly pariahs were received by him on a basis of absolute equality; he refused to make any distinction between queens and courtezans when they came to visit him. During his life-time, at any rate, the Order was a most democratic institution. 'One of the most distinguished members of his Order,' writes Rhys Davids, 'the very one of them who was referred to as the chief authority, after Gautama himself, on the rules of the Order, was Upali, who had formerly been a barber, one of the despised occupations. So Sunita, one of the Brethren whose verses are chosen for insertion in the *Thera Gatha*, was a pukhusa, one of the low tribes. Sati, the propounder of a deadly heresy, was of the sons of the fisherfolk, afterwards a low caste, and even then an occupation, on account of its cruelty, particularly abhorred. Nanda was a cowherd. The two Panthakas were born out of wedlock, to a girl of good family, through intercourse with a slave (and hence were, according to Manu, to be regarded as outcasts). Kapa was the daughter of a deer-stalker. Punna and Punnita had been slave-girls. Sumangalamata was daughter and wife to a worker in rushes, and Subha was the daughter of a slave.' Had Gautama's views been accepted by his countrymen, the Hindu caste system would never have come into existence.

There were many picturesque and interesting personalities amongst Gautama's friends and followers. In high places, there were the kings of Magadha and Kosala. Both Bimbisara and Pasenadi were deeply devoted to the Tathagata, and remained so till they met their death in the tragic circumstances which have been mentioned elsewhere. Bimbisara, in particular, held Gautama in very high esteem. He not only gave valuable gifts to the Brotherhood, but actually found time from his varied kingly duties to visit the Buddha frequently and hear the doctrine.

Gautama's other royal benefactor, Pasenadi of Kosala, was a man of sober-minded bent. He does not seem to have been quite so lively and entertaining a figure as king Bimbisara. He was far less effusive and vehement in protesting his abiding regard for the Blessed One. But probably his interest in the Eightfold Path was more genuine. He had been educated at the university of Takhasila (modern Taxila), the capital of Gandhara. This early training had developed in him a seriousness of purpose and intellectual honesty sadly lacking in the inconsequential, though affable, Bimbisara. He showed himself highly efficient in the discharge of his administrative duties, and his serious philosophic nature made him cultivate 'the companionship of the good.' His admiration for Gautama was based on a true understanding of the man's worth.

Besides these royal personages, Gautama counted among his lay followers many gentlemen of noble birth, aristocratic patriarchs, chiefs of clans, wealthy merchants, and guildmasters. The term 'the friends of the series of the wealthy ones' which occurs in the Sanskrit texts is not without significance. Sudatta, alias Anathapindaka, was not the only man who gave alms 'to the unprotected'; doubtless there were many other guildmasters who were equally charitable towards the Brethren. Further, Gautama's circle was not confined only to those who accepted his doctrine. It included intellectuals, dialecticians, mystics, and philosophers holding views widely different from his own. Indeed, he was perhaps better understood by those who disagreed with him than those who avowed adherence to his doctrine. And although he was up in arms against the Brahmanical orthodoxy and voiced the most devastating heresies, the Brahmans showed on the whole a surprising tolerance towards him and his followers. The more enlightened among them—men like Pokkharadi and Sonadanda for instance—not only respected his point of view, but sympathized with it, though they were far too deeply committed to orthodoxy to come out into the open with their sympathies.

Among the members of the Order itself, Ananda, Sariputta, Moggallana, Upali, and Devadatta enjoyed a certain pre-eminence. Another important personage, though not a member of the Brotherhood, who was on intimate terms with the Tathagata was the physician Givaka Komarabhakha. Givaka had risen to fame and wealth in a most remarkable and adventurous manner. He was the son of a courtezan. His mother had abandoned him immediately after his birth, fearing her worth would sink in the eyes of men if they were to know that she had borne a child. After an unhappy childhood spent in dire poverty and misery, Givaka somehow drifted to Takhasila. Here he was lucky enough to attract the attention of a renowned physician, who so took to the youth that he adopted him as his son, and taught him all that there was to teach about human diseases and their cures. Givaka showed a great aptitude for medicine, and before long was himself an accomplished physician. Then came the chance of his life. 'At that time,' says the Vinaya, 'the Magadha king Seniya Bimbisara suffered from a fistula; his garments were stained with blood.' As though the agony of a fistula were not enough, the ladies of the seraglio took a perverse pleasure in mocking him. When they saw his blood-stained garments, they jeered and said: 'His Majesty is having his courses! His Majesty will bring forth!' Bimbisara was annoyed, but could do nothing to prevent his playful concubines making fun of him, especially as he was already advanced in years. He confessed his misery to the Royal Prince. 'I am suffering, my dear Abhaya,' he moaned, 'from such a disease that my garments are stained with blood; and the queens when they see it mock me. Pray, my dear Abhaya, find me a physician who can cure me.' Abhaya's choice fell on Givaka; and a better choice could not have been made. Givaka performed a successful operation on Bimbisara. As a token of his gratitude, the merry monarch appointed Givaka to the coveted post of Royal Physician. Later, when Bimbisara discovered the virtues of the Eightfold Path and became a patron of the Buddha, he generously placed Givaka's services at the disposal of the Tathagata.

Gautama had many female admirers. He was far too attractive a personality not to have aroused the maternal instinct. We hear of many charitable ladies who contributed to the general ease and happiness of the Brethren in diverse ways. 'The work of ministering to the Order,' writes Sir Charles Eliot, 'of supplying it with food and raiment, naturally fell largely to pious matrons, and their attentive forethought delighted to provide for the monks those comforts which might be accepted, but not asked for.' A notable figure among these ministering ladies was Visakha, wife of

Punnavaddhana, son of Migara. Domestic unhappiness had led her to take refuge in the Buddha as a lay follower. She is said to have built a monastery at Savatthi in the Pubbarama, or 'the Eastern Park,' for the use of the Brotherhood. On another occasion, it is recounted, she asked eight boons of Gautama. These included permission to provide food for various types of monks, and to furnish them with bathing-costumes—'for,' said she, 'it shocks my sense of propriety to see them bathe naked.' Gautama had enough common sense to grant these boons.

XI

FOOLS IN THE ORDER

UNDERSTANDING has never been among the conspicuous virtues of the faithful. There is a very touching passage in the Dhammapada which may well represent Gautama's own heart-felt conviction. 'If a traveller,' it says, 'does not meet with one who is his better, or his equal, let him keep to his solitary journey; there is no companionship with a fool.' For Gautama had to make most of his journey in the company of people who were little better than fools. It was not just the cussedness of his uncomprehending enemies from which he had to suffer; in this respect the members of his own fraternity were probably the worst sinners. He had founded the Brotherhood in the hope of creating a nucleus for a new and more humane kind of society; those joining it were expected to propagate, by the example of their strenuous effort and vigilance as much as by their exhortations, a new and more reasonable way of life. Yet, from the stories recounted in the Vinaya, it would appear that the 'worthy Brethren' thought of almost everything except the purpose which the Buddha had so much at heart.

This was inevitable. The composition of the Order was far too heterogeneous for it to be a body of altruistic humanitarians intent on furthering 'the welfare of mankind.' The motives out of which many people had come to 'disburden' themselves were questionable. Some had taken refuge with the Buddha because it was a new craze. Others had attached themselves to the Order thinking 'the precepts which these Sakyaputtiya Samanas keep and the life they live are commodious; they have good meals and lie down on beds protected from the wind.' There were many more who had renounced the world under the stress of necessity; these included runaway slaves and domestic servants, deserters from the army, debtors in distress, criminals fleeing from Justice, youths wishing to break away from the tyranny of family-life, bankrupt business men, and harassed householders. Indeed, at one period these stricken souls were joining the Order in such alarming numbers, that the authorities got so 'vexed and indignant,' and 'murmured' so effectively, that Gautama had to make the rules for 'Disburdenment' much more strict than they were at the beginning. However, in spite of all the caution that was exercised, it was impossible to prevent 'undesirables' from furtively attaching themselves to the Brotherhood.

Not all the 'worthy Brethren' had the perfect deportment of the elder Assaji, whose 'decorous walk and looks and motions of the arms' had made such a deep impression on Sariputta's mind. In fact, most of them were rather uncouth and noisy fellows who paid little heed to the rules of decorum and decency. Innumerable examples of their bad behaviour are recorded. They left doors and lattices in the monastery open. They left bathrooms in a filthy condition. They behaved rudely towards the householders when they went out to beg alms, and 'entered dwellings roughly and left them roughly.' They wrestled, and held boxing bouts, and rubbed their bodies with oil as though they were going to take part in some physical culture championship. They ate their meals noisily and carelessly. They walked on the couches with unwashed and dirty feet. They broke furniture belonging to the Brotherhood to light bonfires. Some rose up in the night towards dawn, 'and putting on wooden shoes, walked up and down in the open air, talking in tones high, loud and harsh . . .; and in so doing they both trod upon and slew all kinds of insects and disturbed Bhikhus in their meditation.' The dead insects, of course, could not protest; but the Bhikhus, who were disturbed in their meditation, would get 'vexed and indignant.'

The world-renouncers often succumbed to the lure of the world. They would get heavily drunk and then lie about the monastery, unmindful, and displaying their nakedness. The number of serious misdemeanours increased vastly when the Order of the Female Mendicants was brought into being. Gautama's reluctance to take this step would seem to have been justified; for bhikhus and bhikhunis frequently misconducted themselves together. On one occasion, we read, some of the Brethren had the original idea of wooing some of the bhikhunis by throwing dirty water at them. Bhikhunis, on their part did everything 'to set men's hearts ablaze,' and would bedeck themselves with frills, fringes, and girdles 'as do women who are still enjoying the pleasures of the world.' Besides such cases of misconduct, the living together of men in comparative segregation from women, led to a certain amount of homosexuality.

The 'temperate' bhikhus had good reasons for being annoyed. Their fellow-mendicants were always doing something or other which brought discredit either on themselves or the Order. Of course, the bhikhus who caused trouble were not always moved by the spirit of evil; quite often their offence was merely due to an incorrigible naïvety of temperament. A majority of the Brethren were simple-minded people who knew little of the ways of the world, and consequently found the Middle Way even harder to follow. Especially, they seem to have been very prone to fits of absent-mindedness. Some of them suffered from this malady so much that, when going out to beg alms, they forgot to put on proper robes and under-garments, and aroused the people's hostility or amusement.

Not much love was lost between individual members of

the Brotherhood. Factions, rivalries, petty jealousies and quarrels were rife, and not all the sweet reasonableness of the Tathagata could convince his cantankerous disciples of the futility of trying 'to appease hatred by hatred.' Sometimes the enmity between Brethren broke out in a virulent form; then the contending parties would not scruple to level charges and counter-charges of the most serious kind. In particular, there was constant friction between the novices and the elders. The former, we read, 'did not show reverence and confidence towards the Bhikhus, and did not live in harmony with them. The Bhikhus were annoyed, murmured, and became angry.' The Bhikhus, on their part, took every chance to pin-prick the novices and impose hardships on them, with the result that novices also were often 'annoyed, murmured, and became angry.'

Gautama himself was possessed of a quality of compassion such as has seldom been seen among men. His sympathy was all-embracing and spontaneous; his generosity was equalled only by his profound understanding of human nature. When Panthaka came to him homeless and weary, Gautama tended him with his own hands. It is best to give the description of the incident in Panthaka's own words. 'Then the lord,' he says, 'came and stroked my head and taking my arm led me into the garden of the monastery, and out of his kindness he gave me a towel for my feet.' Another time he refused to preach until a farmer who had come to hear his discourse and had not eaten, had been fed.

But Gautama's example of gentleness and pervasive love appears to have been lost on most of his followers. For instance, when one of the bhikhus was suffering from an offensive skin disease and lay in his own excreta, none of the Brethren would go near him. Gautama saw him lying untended and was deeply grieved. He had him carried to the monastery, heated water for his bath, and himself bathed him. Then when the diseased mendicant had been clothed in dry and freshly washed garments, Gautama sadly told the Brethren: 'You, monks, have no mother or father. If you do not wait one on the other, who is there who will wait on you? Whosoever, monks, should wait on me, he should wait on the sick.' But such exhortations, though they might have caused a moment's remorse, apparently did not leave any enduring impression on the monks' minds. The gods themselves, says Schiller, struggle in vain against stupidity. The Tathagata's struggle with *Dummheit*—or rather, DUMMHEIT—seems to have been equally fruitless.

On reading the section in the Vinaya which deals with rules concerning footwear, clothing, etc., one is perplexed by the utter incapacity of the Brethren to understand even the most elementary points of discipline. If so many misunderstandings could arise on such minor issues, it is hardly surprising that there should have been gross misinterpretations and distortions of the doctrine. Not all the members of the Order were experienced dialecticians. In fact, their general level of intelligence appears to have been astonishingly low; even the best among them exhibited a distressing half-wittedness.

It was only a strong sense of humour that saved Gautama from frequent heartbreak. His irony was so gentle and effective that a few examples might be quoted. Once, towards the end of his life when Gautama was staying at Nalanda in the Mango grove of Pavarika, Sariputta, who was famous for his understanding of the doctrine, and was even called 'the Spotless Eye of the Doctrine,' came and said to him: 'Lord! Such faith have I in the Exalted One, that methinks there never has been, nor will there be, nor is there now, any other, whether recluse or Brahman, who is greater and wiser than the Exalted One, that is to say as regards the higher wisdom.' Gautama, who always tried to discourage the tendency to hyperbole, replied: 'Grand and bold are the words of thy mouth, Sariputta. Thou hast roared a veritable lion's roar in this that thou hast said. Of course, then thou hast known all the Exalted Ones who in the long ages of the past have been Arahats, Awakened Ones, comprehending their minds with thy mind, and aware what was their conduct, what their doctrine, what their wisdom, what their mode of life, and the liberty to which they attained?' Sariputta, amazed that his praise had not been appreciated, replied: 'Not so, lord.' 'Of course,' continued the Buddha, 'then thou hast perceived all the Exalted Ones who in the long ages of the future will be Arahats, Awakened Ones, comprehending their minds with thy mind, and aware what will be their conduct . . . doctrine . . . wisdom . . . mode of life and . . . liberty?' 'Not so, lord,' confessed the wondering Spotless Eye of the Doctrine. 'But at least then, Sariputta,' continued the Buddha, 'thou knowest that I now am Arahat, an Awakened One, comprehending my mind with thy mind, and aware that this is the Exalted One's conduct, such is his wisdom, such his doctrine, such his mode of life, and such the liberty to which he has attained?' 'Not so, lord,' blurted out Sariputta in confusion. 'Lo! then, Sariputta,' concluded the Buddha driving his point home, 'no knowledge hast thou concerning Arahats, Awakened Ones, past, future or present. Why then forsooth are thy words so grand and bold? Why hast thou roared this all-comprehensive lion's roar?'

When Ajatasattu, after having starved his father to death and usurped the throne, was stricken with remorse, he came to Gautama and confessed his transgression, saying, 'transgression overcame me, lord, in that in folly, stupidity, and wickedness, for the sake of lordship I deprived my righteous father, the righteous king, of life. May the Lord accept my transgression as transgression that I may be restrained in the future.' The lord, of course, accepted Ajatasattu's transgression—there was no alternative—but he could not refrain from commenting that 'His Majesty had all the makings of a saint in him, if only he had not killed that excellent man, his father.'

Gautama was often pestered by people who were desirous of knowing the fate of their departed dead, and hearing stories of the Blessed One's own adventures in his previous births. Out of politeness Gautama is said to have complied with these requests, though, it is added, he always had an enigmatic smile on his face while telling these fairy-tales. But there were moments when his patience almost gave out. Thus when Ananda came to him and said: 'The brother named Salha has died at Nadika, lord. Where has he been born, and what is his destiny? The sister named Nanda has died, lord, at Nadika. Where is she born, and what is her destiny?' and went on to enquire after a dozen other departed souls, Gautama was really rather unhappy. 'Now there is nothing strange in this, Ananda,' he sadly answered his credulous disciple, 'that a human being should die, but that as each one does so, you should come to the Buddha, and enquire about him in this manner, that is wearisome to the Buddha.'

For all his sense of humour, his subtle irony, his patient nature, and his mature wisdom, the Tathagata must have been at heart a sad man.

DEATH OF BUDDHA

(Swat—*Copyright, Archaeological Survey of India*)

XII

THE GREAT DECEASE

Such was the end of our friend, whom I may truly call the wisest, the justest, and best of all men whom I have ever known.

Plato in *Phaedo*

THE day's journey was drawn out into a long and wistful twilight. It was a troubled twilight.

In the first place, the political horizon in India during the closing years of Gautama's ministry seems to have been anything but peaceful. Oldenberg writing in the late nineteenth century could, of course, afford to contemplate the remote Buddhist times with perfect serenity, and say: 'For those circles of the Indian race among whom Buddha preached his doctrine the idea of non-Indian lands had hardly a more concrete signification than the conception of those other worlds, which, scattered through infinite space, combined with other suns, other moons, and other hells, to form other universes.' But political circles in Gautama's days were not in a position to take this complacent view of the situation confronting them. Unfortunately, these non-Indian lands had already acquired a painfully concrete and immediate actuality.

While Gautama was harping on the impermanency of things and preaching his gospel of universal compassion, the restless rulers of one of the other countries had been very differently occupied. The ruthless and overwhelming armies of Cyrus, and his grandson Darius, had covered not only Asia, but parts of Africa and Europe 'with blood and flames.' Though, of course, the Persian monarchs could not claim that the sun never set on their Empire, they could, and did, boast 'that they ruled from the land of uninhabitable heat to the land of uninhabitable cold; that their dominion began in regions where the sun frizzled the hair and blackened the faces of the natives, and ended in a land where the air was filled with snow like feathers, and the earth was hard as stone.' Nor was this merely a boast: it was an imaginative description of facts. The Great Kings of Glory apparently did not spend all their time in attending to their international harem where, says Winwood Reade, 'there might be seen the fair Circassian, with cheeks like the apple in its rosy bloom; and the Abyssinian damsel, with warm brown skin and voluptuous drowsy eyes; the Hindoo girl, with lithe and undulating form, and fingers which seemed created to caress; the Syrian, with aquiline and haughty look; the Greek, with features brightened by intellect and vivacity; and the home-born beauty prepared expressly for the harem, with a complexion as white as the milk on which she had been fed, and a face in form and expression resembling the full moon.' Their practical Zoroastrian wisdom enabled them to combine pleasure with empire-building.

Thus they had created an Empire the like of which had never been seen on the earth before, and which an appreciative Greek historian thought would last for all eternity; an Empire 'bounded by the deserts which divided Egypt from Ethiopia on the south, and from Carthage on the west . . . by the steppes which lay on the other side of the Jaxartes; by the Mediterranean, the Caspian, and the Black Sea.' And more! The Persian expansion was not confined to the western regions. The Great Kings of Glory were equally susceptible to the lure of the dawn; they realized the immense possibilities of expansion in the direction of the Rising Sun; and the conquest of the Indo-Gangetic plain was as much a part of their policy as the subjugation of lands watered by the Tigris, the Euphrates, the Nile, and the Danube. The Behistun inscription, it is true, says nothing about India; but from the Persepolis and Nakhs-i-Rustam inscriptions it may safely be inferred that 'the country of the Indus' figured prominently on the plans of the Persian General Staff. The *Kyropaedia* of Xenophon credits Cyrus with the conquest of Gandhara, a province consisting of the modern Peshawar and Swat river districts. Herodotus confirms this, adding that Darius extended his Indian possessions to the lower Indus valley—that is from Kalabag to the sea. And it is certain that when the eventful sixth century drew to its close, Gandhara was the richest province of the Persian Empire and paid a regular yearly tribute in gold-dust valued at 4,680 Euboic talents—equivalent to about one million pounds sterling.

Indian political circles were neither deaf nor blind. Trade-relations between Persia and India had existed from time immemorial. The efficient system of inter-Imperial highways worked out by the Persians had vastly improved the trade between the two countries. Traders and many other adventurers brought to the Indian princes news of the exploits of the Great Kings of Glory. And these Indian rulers had heard also of a mysterious flotilla commanded by a Carrian sailor which had sailed down the Indus, along the tortuous coastline of the Arabian sea, cut across the straits of Ormuz, through Bab-el-Mandeb into the Red Sea, finally shoring in Egypt after a voyage lasting two and a half years, in the course of which Skylax and his fearless crew sailed past deaths and hazards such as Ulysses never knew. Moreover, the ruling powers in India were aware that the king of Gandhara was a king in name only, and his dominion a mere satrapy of the Persian Empire. All these factors combined to make them anxious. Persia, though a 'non-Indian land,' seemed to them dangerously near; unlike Oldenberg, they could not afford to see it as a planet in some timelessly remote solar system, 'which, scattered through infinite space, combined with other suns, other moons, and other hells, to form other universes.'

But the Great Kings of Glory were not merely a menace; they were also a highly inspiring example. The ambitious princes in India could not but envy their achievements. If Cyrus, who, after all, had been only the chieftain of an obscure and uncouth highland clan in the land of Zoroaster, could rise to such deeds, why should they not achieve as much? Indeed, they could hardly help feeling that they were destined for even higher things, since some of them could trace back their lineage to the sun and the moon.

Thus began the struggle for political domination in India. It necessarily involved much bloodshed. Supremacy could only be won by exterminating rival powers. And so, paradoxically enough, when Gautama was preaching peace and goodwill, the ambitious princes were brooding over their plans for exterminating each other. Gautama was to live long enough to see them put into execution; long enough to witness the beginnings of monarchical reaction and the disappearance of aristocratic republicanism as a political force in India.

Of course, the political pattern of the country had been changing throughout the duration of his ministry; but during the later years the process was appreciably accelerated. The change was particularly obvious in those parts of India to which Gautama's wanderings were mainly confined. Things were moving swiftly towards a crisis in the Gangetic valley. Republican confederations like those of the Vajjians, the Licchavis, and the Videhans seemed doomed to extinction, being far too inchoate and divided to withstand attacks by war-like princes. Some of the minor republics had already become extinct. And it was clear to all those who had any political sense that, as far as the sacred valley of the Ganges was concerned, the question of supremacy must eventually be fought out between the rulers of Magadha and Kosala.

Both in Magadha and Kosala youth was at the helm; Ajatasattu in Magadha, and Vidudabha in Kosala. And Gautama was now an old man. He found it difficult to understand the impetuous ways of youth or adjust himself to them. Moreover, youth had come to power by methods which the Tathagata could hardly have approved. Both Ajatasattu and Vidudabha had usurped the thrones from their 'righteous fathers.' Though Ajatasattu had starved his father, Bimbisara, to death, fate had saved Vidudabha from becoming a patricide; for Pasenadi, while on his way to Ajatasattu to seek help against his treacherous son, had died of exposure. Both Bimbisara and Pasenadi had been Gautama's friends and patrons; he could not help feeling their deaths as an irredeemable personal loss. It is true, that when Ajatasattu came and asked forgiveness, Gautama forgave him, but his ironical remark at the interview would indicate that he did not place much value on the penitential confession of the remorse-stricken monarch of Magadha.

There was more tragedy ahead. And the light was fading. Vidudabha, unlike Ajatasattu, was not subject to pangs of remorse. He never appeared in sackcloth and ashes before the Tathagata to ask forgiveness. On the contrary, it was the Tathagata who had to go to him to supplicate on behalf of his kinsmen. For, soon after having come to power, Vidudabha began to make plans for expanding his kingdom. These included the annexation of the Sakya territory. Indeed, it was his immediate objective. He knew that on the other side Ajatasattu was engaged in fortifying his frontier in the direction of the Vajjian confederacy, and had in fact built a fort at Pataligama—then an insignificant village on the Ganges, which was destined to become in less than two centuries the capital of the great Maurya Empire. He knew that conquest of the Sakya republic would place him in an ideal strategic position not only to strike at the Vajjians before the king of Magadha had time to complete his fortifications, but to attack Magadha itself if the stars were favourable to his adventure. Vidudabha's plans somehow became known to Gautama. Seriously disturbed about the imminent fate of his clan, he went in person to the ambitious king of Kosala and tried to induce him to spare the Sakyas. But arguments and supplications were wasted on the obstinate Vidudabha. Three years before Gautama's death, Vidudabha attacked Kapilavastu, perpetrated a most horrible massacre of the Sakyas, sparing neither children nor women, and razed the township to the ground. The reason given for Vidudabha's violence where the people of Kapilavastu were concerned was that he bore the Sakyas a personal grudge. The Sakyas had tricked his father into marrying Mahanama, the daughter of a Sakya by a slave-woman. This girl was Vidudabha's mother; and Vidudabha while on a visit to his mother's home had accidentally discovered his mother's base origin, and therefore his own. The stigma of low birth was hard to bear, and he vowed to avenge himself on the Sakyas at the first available opportunity. How far this is true it is difficult to say. But the annexation of the Sakya territory by Vidudabha is probably historical fact; and the wrathful manner in which it was carried out would certainly suggest that, apart from political motives, Vidudabha was actuated by a sense of personal grievance. In any case, whatever the nature of his motives, the destruction of Kapilavastu and the ruthless slaughter of its people were not likely to contribute to the sum of the Tathagata's happiness during his twilight days.

But not all the misfortunes of the Tathagata were the outcome of political unrest. He had trouble with his own recalcitrant followers. There was the incorrigible Devadatta. He had always taken a perverse pleasure in making things difficult for the Buddha. During the later years of Gautama's ministry he became increasingly insolent and hostile. He was said to have been at the back of Ajatasattu's conspiracy to depose his father from the throne and starve him to death. Once when Gautama was spending the Retreat at Rajagaha, he dared to get up while Gautama was delivering a sermon and say: 'The Blessed One, lord, is now grown aged, he is old and stricken in years, he has accomplished a long journey, and his term of life is nearly done. Let the Blessed One now dwell at ease in the enjoyment of happiness reached even in the world, let the Blessed One give up the Bhikhu Brotherhood to me.'

For obvious reasons, Gautama was not prepared to accept Devadatta's embarrassing suggestion. This enraged Devadatta and he left the Order in anger. So far Gautama had behaved with utmost patience towards him; even when Moggallana had told him about the part played by Devadatta in bringing about Bimbisara's death, he had merely answered that it was better not to mention such things and let the foolish man reveal himself. However, it now became necessary to safeguard the Order against Devadatta. A public warning was therefore given, to the effect that in future in anything that Devadatta might do, he should be regarded as acting in his individual capacity and not as a member of the Brotherhood.

This was the first open defection from the Order. Evidently, however, even after the above proclamation had been issued, Devadatta continued to regard himself as a Buddhist. Indeed, he claimed that he was a better Buddhist than the Buddha. To prove which claim in the eyes of the world, he went to Gautama and demanded that certain new and more strict rules should be inaugurated. These rules would have made it incumbent upon the monks to spend all their lives in a forest at the foot of a tree, to live only on begged food and refuse all invitations to meals, to wear cast-off rags, and to abstain from fish and flesh. Gautama was not prepared to promulgate these stringent regulations, though he had no objection to their being kept by monks who so desired. For himself he did not wish to enforce any needlessly harsh ascetic discipline. The higher life, he had always maintained, was not contingent on dwelling in a forest, sitting cross-legged under a tree, eating begged food, or wearing cast-off garments. If so eminent a sage as Yajnavalkya could attain perfect wisdom even after openly declaring that 'he for one ate beef, provided it was tender,' there was no reason why he should inflict vegetarianism on the Brethren. The higher life did not depend on being a vegetarian any more than on lying upon a bed of nails; and he for one ate whatever was given him by the people, no matter whether it was fish, fowl, or carrots. All this he explained to Devadatta when refusing his suggestion.

A refusal was precisely what Devadatta wanted. He could now proclaim to the world that the Buddha was a fraud, that he was leading a self-indulgent life. By way of contrast, and to show how much more religious he himself was, he founded an Order in which it was compulsory for the members to observe the discipline which he had recommended to Gautama. For a while his society flourished. It was supported by no less a man than Ajatasattu. But death seems to have cut short his career as a religious teacher, and his influence naturally declined, even Ajatasattu becoming—nominally at least—a follower of the Buddha. However, the schismatic growth of which he was the cause did not apparently end with him; when the Chinese pilgrim, Fa

Hsian, visited India in the fifth century A.D. he found that the cult of Devadatta was still in vogue.

The light was failing. And so was Gautama's health. Even the digestive organs of a Buddha are subject to the decay inherent in all component things. Gautama's health had indeed never been very robust all through the long years of his ministry. The severe penances he had undertaken had not had a particularly happy effect on his intestines. For the rest of his life he had periodic trouble from this source; a trouble that grew frequent and acute with age. Givaka Komarabhakha, the Physician-Royal, did what he could to mitigate the Tathagata's suffering. But there were times when even his excellent physic brought no relief. Some idea of Gautama's agony may be formed from the fact recorded in the Vinaya, that on one occasion Givaka had to administer to him thirty successive doses of a strong purgative before the Tathagata's bowels would move.

Gautama had grown very old. In the twilight of age certitudes have a habit of falling away, of vanishing. It is one thing to reject other-worldly consolations and describe the Deity as an irrelevance when a man is thirty-five, but to do so at eighty is quite another matter altogether. And Gautama was now almost eighty. He was no longer a person; he had become a kind of institution. But an institution from the past. Most of his own generation, people he had known and liked, were no longer alive. Another race had been born; other palms were being won. Pasenadi and Bimbisara had long been dead. Sariputta and Moggallana also, the latter having been beaten to death by a band of robbers. Even that irritant, Devadatta, was no longer there to create trouble. But the old Buddha lived on; and sometimes it almost seemed as if he would go on living till eternity. Even his own disciples could hardly help feeling that it was time for the Tathagata to cross to the shores of the Great Beyond, to attain Nirvana. In fact, when a few months before his death, Gautama told Ananda that he could, if he wished, prolong his life indefinitely, it did not at first occur to Ananda to request the Tathagata to do so; and seeing his faithful disciple's unconcern, Gautama remarked with his usual irony: 'Make thyself happy, the final extinction of the Tathagata will take place before long.' Ananda began to protest, half-heartedly pleading that the Tathagata should live on for an epoch. But it was too late. Gautama, still gently ironical, reminded Ananda that on sixteen previous occasions he had dropped similar hints and they had not been taken by him. 'If, Ananda, you had asked the Tathagata,' he added, 'he might have refused twice, but he would have assented the third time . . . Enough now, Ananda, beseech not the Tathagata. The time for making such request is past.' In his bewilderment and shame Ananda could scarcely say a word. The Tathagata could be baffling even at eighty.

The end came at last. It was overdue. The account of Gautama's last days is given at considerable length in the *Book of the Great Decease*. The story is needlessly loaded with miracles and prodigies, literally clogged with supernatural details. Nevertheless, it contains passages of moving beauty; passages over which there broods a strangely haunting melancholy and pathos. We are given the picture of a man weary and overwhelmed with the burden of age; yet a man who, for all his physical misery, retained to the last a sense of proportion, a serene dispassionateness, an uncompromising intellectual honesty—and a tenderness for life that is not to be grasped by 'mere logic.'

The *Book of the Great Decease* covers a period of about three months, and describes Gautama's last journey from

Rajagaha, through the valley of the Ganges to Vesali, and thence by degrees to Kusinara, a minor township in the territory of the Mallas. Gautama had spent the forty-fourth rainy season of his ministry at Savatthi, in the monastery of Jetavana. From there he returned to Rajagaha and dwelt for a time on the hill called the Vulture's Peak, which looked down upon the beautiful valley of Rajagaha. The city was then full of rumours about the plans which Ajatasattu was making to attack the Vajjian republic; indeed the ambitious monarch had himself declared, 'I will root out these Vajjians, mighty and powerful though they be, I will destroy these Vajjians, I will bring these Vajjians to utter ruin.'

Whether it was to warn the Vajjians of the impending attack, or for some private reasons, Gautama decided to go to Vesali. He crossed the Ganges near Pataligama, where Ajatasattu was building his new fort in preparation for his attack, and reached Vesali after halting at a few wayside villages. Here he annoyed the Vajjian nobles by accepting the invitation of Ambapalika, the famous courtezan through whom the township had become 'opulent, prosperous, populous, crowded with people, abundant with food . . .'

Here he also delivered one of his best later discourses, in which he said: 'Herein, O mendicants, let a brother, as he dwells in the body, so regard the body that he, being strenuous, thoughtful, and mindful may, while in the world, overcome the grief which arises from bodily craving—while subject to sensations, let him continue so to regard the sensations that he, being strenuous, thoughtful, and mindful, may, whilst in the world, overcome the grief which arises from the craving which follows our sensation—and so also as he thinks or reasons or feels let him overcome the grief which arises from the craving due to ideas, or reasoning, or feeling.'

From Vesali he proceeded to Beluva. The rains had now set in, and Gautama decided to keep Retreat in that village. During this Retreat 'there fell upon him a dire sickness, and sharp pains came upon him, even unto death.' His disciples despaired of his life and were apprehensive that 'the Blessed One should pass away without having left instructions as touching the Order.' But much to everybody's surprise Gautama recovered from his illness. As soon as he had regained enough strength, Ananda went to him and asked him about his final instructions. Gautama's reply was characteristic. 'What, then, Ananda?' he asked. 'Does the Order expect that of me? I have preached the truth without making any distinction between exoteric and esoteric doctrine: for in respect of the truths, Ananda, the Tathagata has no such thing as the closed fist of a teacher, who keeps some things back. Surely, Ananda, should there be any one who harbours the thought, "It is I who will lead the Brotherhood," or "The Order is dependent upon me," it is he who should lay down instructions in any matter concerning the Order. Now the Tathagata, Ananda, thinks not that it is he who should lead the Brotherhood, or that the Order is dependent upon him. Why then should he leave instructions in any matter concerning the Order?' By way of a hint that he should not be pestered with stupid questions now that he was old and weary, he said: 'I too, O Ananda, am now grown old, and full of years, my journey is drawing near to its close, I have reached my sum of days, I am turning eighty years of age; and even as a worn-out cart, Ananda, can only be kept going with much additional care, so, methinks, the body of the Tathagata can be made to move along only with much additional care . . .' And he ended by saying: 'Therefore, O Ananda, be ye lamps unto yourselves. Be ye a refuge to yourselves. Betake yourselves to no external refuge. Hold fast to the truth as a lamp. Hold fast as a refuge to the truth. Look not for refuge to any one besides yourselves . . . And whosoever, Ananda, either now or after I am dead, shall be a lamp unto themselves, and a refuge unto themselves, shall betake themselves to no external refuge, but holding fast to the truth as their lamp, and holding fast as their refuge to the truth, shall look not for refuge to any one besides themselves—it is they, Ananda, among my bhikhus who shall reach the very topmost height—but they must be anxious to learn.' Evidently, despite old age and the doubts and despair of old age, despite the merciless worm of decay, despite even the intestinal agony, the light had not yet altogether failed the Tathagata. The twilight vision had somehow retained the lucidity of dawn.

After the rains Gautama made a slow tour of the villages in the neighbourhood of Vesali. He knew that he could not live long, and therefore wanted to visit for the last time the people and places he had known. It was autumn now. What had been jade was now gold. Then the leaves fell; and gold was mingled with dust. The Tathagata's mind lingered over all that he had seen and felt and known. In the nostalgic vision, the world of transient things, the world of suffering and pain, was transfigured into something magically beautiful, and desirable. Again and again Gautama would remember the name of some place and say to his companion, 'Ananda how pleasant it was!' It was all pleasant—even the pain.

By slow stages he reached Pava. Here he was entertained by Kunda, a worker in metals. Kunda prepared dried boar's flesh, sweet rice, and cakes for the Tathagata's dinner. The food was perhaps too rich and solid for the Tathagata's stomach; or perhaps the meat was bad. For when Gautama had eaten it 'there fell upon him a dire sickness, the disease of dysentery, and sharp pain came upon him, even unto death.' However, Gautama bore his suffering 'without complaint and with fortitude.' He was even rash enough to start for Kusinara, a town in the country of the Mallas. Half-way to Kusinara he felt completely exhausted, and going aside from the road to the foot of a certain tree, he said to his companion: 'Fold, I pray you, Ananda, the robe; and spread it out for me. I am weary, Ananda, and must rest awhile.' Ananda did as he was told. Then Gautama asked for some water, which Ananda brought from the river nearby. After having drunk some water he bathed in the river. This gave him sufficient energy to continue his journey to Kusinara. As usual, he stayed in a grove outside the town. Here he told Ananda that after his death they should not blame Kunda for having been the cause of it. He also began giving instructions as to how his remains were to be disposed.

In the middle of the conversation Ananda broke down. He went aside to weep. He had been attached to Gautama all these years with a kind of filial devotion, and could not bear the thought of his passing away. Gautama heard him sobbing as he stood leaning against the lintel of the door. He called his disconsolate disciple and tried to console him. 'Enough, Ananda,' he said to him, 'do not let yourself be troubled; do not weep. Have I not already, on former occasions, told you that it is in the very nature of things most near and dear unto us that we must divide ourselves from them, leave them, sever ourselves from them? How, then, Ananda, can this be possible—whereas anything whatever born, brought into being, and organized, contains within itself the inherent necessity of dissolution—how, then, can this be possible, that such a being should not be dissolved? No such condition can exist! For a long time, Ananda, have you been very near to me by acts of love, kind and good, that never varies and is beyond all measure. You have done well, Ananda! Be earnest in effort, and you too shall soon be free from the great cankers—from sensuality, from individuality, from delusion, and from ignorance.'

Ananda was consoled and was soon making the naïve suggestion that the Tathagata should not die in a cheap little town like Kusinara, but should choose some big city such as Benares, Savatthi, or Kosambi 'where there are many wealthy nobles and Brahmans and heads of families, believers in the Tathagata, who will pay due honour to the remains of the Tathagata.' Gautama did not give much heed to this suggestion. Instead, he asked Ananda to go to the town and tell the people that his end was near. Ananda put on his robes and went.

The night had fallen. The Tathagata lay on his bed in a state of semi-consciousness, awaiting the moment of final disburdenment. During the first watch of the night a large number of the inhabitants of Kusinara came to see the dying Tathagata. Later in the night, Subhadda, a mendicant of the place, who was not a believer, asked to speak to Gautama in order to have his doubts resolved. But Ananda stopped him at the door, saying: 'Enough, friend Subhadda! Trouble not the Tathagata. The Blessed One is weary.' Yet Subhadda persisted in his request. The noise of the altercations between these two was overheard by Gautama, and he asked Ananda to let Subhadda come in.

Subhadda took his seat and began, 'The Brahmans by saintliness, Gautama, are heads of companies of disciples and students, teachers of students, well known, renowned, founders of schools of doctrine, esteemed as good men by the multitude—to wit, Purana Kassapa, Makkhali of the cattle-pen, Ajita of the garments of hair, Kakkayana of the Pakudha-tree, Sanjiya the son of the Belatthi slave-girl, and Nigantha of the Natha clan—have they all, according to their own assertions, thoroughly understood things? Or have they not? Or are there some of them who have understood, and some who have not?' But Gautama had neither time nor strength to go into irrelevant matters. He cut Subhadda short, saying: 'Enough, Subhadda! Let this matter rest whether they according to their own assertions have thoroughly understood things, or whether they have not, or whether some of them have understood and some have not . . .' Then he went on to give briefly the cardinal points of his own doctrine; and Subhadda, hardly understanding, seemed to nod his assent to Gautama's propositions.

The darkness was now fast closing in. But a few rays of light still lingered on the outer edge of night. In a lucid moment he told Ananda that the Order might, if it should so wish, abolish all the lesser and minor precepts. All his life he had fought against the tyranny of 'mere morality,' of convention, of empty piety and form; and he did not want the Brotherhood to degenerate into a snug refuge for such worthies among men as were more interested in holding on to their prejudices than in discovering truth. Finally, he invited all mendicants present to ask him questions about any matters concerning which they had doubts. 'Enquire, Brethren, freely;' he offered, 'do not have to reproach yourselves with the thought, "Our teacher was face to face with us, and we could not bring ourselves to enquire of him when we were face to face with him." ' He offered the suggestion three times, but none of the Brethren came forward. Exhausted, Gautama then sank into a coma. There was, however, one more flicker of light, in the duration of which Gautama said to the mendicants standing round his death-bed: 'Brethren, I impress upon you, decay is inherent in all component things! Work out your salvation with diligence.' These were his last words.

XIII

POSTHUMOUS POSTSCRIPT

GAUTAMA had died in a state of coma. This led to some discussion among the Brethren. Some of them maintained he was dead; others held that he had 'entered into that state in which both sensations and ideas cease to be.' But as soon as they realized that 'the Light of the World had been blown out by the Wind of Impermanency,' they broke forth into loud lamentations. 'Too soon has the Blessed One died,' they bewailed in their bereavement, 'Too soon has the Happy One passed away from existence! Too soon has the Light gone out in the world!' However, there was one man among them who showed no signs of grief. It was Subhadda, the last of the Arahats. He actually argued that the Tathagata's passing away was an occasion for rejoicing rather than sorrow. 'Enough, Brethren!' he was to declare openly a few days later, 'weep not, neither lament. We are well rid of the great Samana. We used to be annoyed by being told, "This beseems you, this beseems you not." But now we shall be able to do whatever we like; and what we do not like, that we shall not have to do.'

The funeral took place on the seventh day after the death. Meanwhile, the news of the Great Decease had spread to the surrounding country. 'Now the king of Magadha, Ajatasattu . . . heard the news that the Blessed One had died at Kusinara,' and he sent a messenger to the Mallas, saying 'The Blessed One belonged to the soldier-caste, and I too am of the soldier-caste. I am worthy to receive a portion of the relics of the Blessed One. Over the remains of the Blessed One will I put up a sacred cairn, and in their honour will I celebrate a feast.'

And the Licchavis of Vesali, and the Sakyas of Kapilavastu, and the Bulis of Allakappa, and the Koliyas of Ramagama, and the Brahman of Vethadipa, and the Mallas of Pava also sent messengers with similar demands. Now that the Tathagata was dead, they all wanted to earn merit by putting up a sacred cairn on his remains and celebrating a feast in their honour. But the Mallas of Kusinara were not prepared to portion out the remains. 'The Blessed One,' they argued, 'died in our village domain. We will not give away any part of the remains of the Blessed One.' This led to angry words between the emissaries of the various clans and the Mallas of Kusinara. Indeed, blood would have been shed over Gautama's ashes had not Dona, a wise Brahman, successfully intervened. He said:

> Hear, reverend sirs, one single word from me.
> Forbearance was our Buddha wont to teach.
> Unseemly is it that over the division
> Of the remains of him who was the best of beings
> Strife should arise, and wounds and wars!
> Let us all, sirs, with one accord unite
> In friendly harmony to make eight portions.
> Wide spread let the Thupas rise in every land
> That in the Enlightened One mankind may trust!

This compromise was accepted. And the Mallas of Kusinara, the Sakyas of Kapilavastu, the Licchavis of Vesali, the Mallas of Pava, the Bulis of Allakappa, the Koliyas of Ramagama, the Brahman of Vethadipa, and the king of Magadha each got their portion of the Tathagata's remains. And the Moriyas of Pipphalivana whose messenger had arrived after the division had taken place were given the embers. And Dona, the wise arbiter, took the vessel in which the body had been burnt. And in ten different places sacred cairns were built, feasts celebrated. Then at last, for a while, there was peace for the Tathagata.

PART III

THE WORD OF THE BUDDHA

In the beginning was the word . . .
St John

I

THE WHEEL

THE world we live in is a world of pain. The inexorable process of life and death, the process of becoming and ceasing-to-be, is profoundly bound up with suffering; at the root of it there is an ever-throbbing sense of grief, of deficit and desolation, like a thorn in the flesh. This, in essence, is the 'noble' truth of pain.

It is obviously not a pleasant truth; by the very nature of things it cannot be a pleasant truth. Yet Gautama took this tragic affirmation as the starting-point of his critique of life. He took it for his starting-point because he was, above all other things, concerned with analysis of the content of human experience, with understanding the elusive pattern of man's subjective universe. His interpretation remains essentially a psychological interpretation. While the metaphysicians of his age, the amateurs as well as the professionals, were wrangling for all they were worth over abstract problems, Gautama seems to have been content with focusing his whole attention on psychological observations. These observations led him to a most disconcerting discovery: the discovery that pain is the most universal, the most significant element in human awareness. The discovery was not, in itself, particularly original; what was original was the emphasis which Gautama placed on it and the comprehensiveness which he gave to it. He made the truth of pain the basis of his doctrine. It has to be accepted in entirety or rejected in entirety; there is no possible middle way. If we reject it, the whole structure of his word crumbles.

It is, of course, possible to reject the truth of pain, to deny its distressing implications. There were many among his own contemporaries who totally rejected his tragic basis. There were the Lokayatikas, or the Worldly-Wise, for instance. These flamboyant philosophers not only denied validity to the 'noble' truth of pain, but they made uproarious fun of the Tathagata. Being worldly-wise, they were impatient of tragic philosophies which might, through slow contagion, undermine their solid certitudes. 'It is absurd,' mocked these hilarious positivists, 'absurd to condemn pleasures because they are mixed with sorrow and dissatisfaction; absurd to throw away rice because the kernel is wrapped in a rough shell.' But their mockery was scarcely a cogent argument; the fact that they thought it to be an adequate answer to Gautama's affirmation was a confession of intellectual confusion. The analogy of rice is misleading. It does not disprove the truth of pain: it merely begs the question. The implication of Gautama's utterance is not that the wrapping is rough, but that the kernel is bitter. He would probably have agreed that the shell of human experience is both soft and tempting, that so long as one is content with the shell one can afford to be oblivious of the reality of suffering. The actuality of life reveals itself, he would have argued, when one has torn asunder the wrapping, penetrated a little deeper than the shell. It is then that an abyss, an apparently bottomless abyss, opens before one; it is then that one is able to recognize 'the misery in the world as it is' in its stark and terrifying actuality.

Whether the implications of Gautama's affirmation are true or not, is an issue which cannot be decided by means of syllogisms. A statement concerning the nature of human experience is not capable of being logically proved or contradicted. The psychological universe is by no means an Euclidean universe: a psychological critique of life, therefore, falls outside the grasp of 'mere logic.' The crucial experience of suffering, even of purely physical suffering—if such a thing is conceivable—is not logically demonstrable. What is more, it eludes all attempts to bring it within the compass of objective comprehension. Ultimately, an aching stomach and an aching soul are equally beyond the sphere of objective understanding: one can dissect the one, and one can record the symptoms of the other, but the immediate actuality of experience in either case cannot be grasped. Strictly speaking, the world of pain is an impalpable and invisible world.

The proper study of human experience, then, is human experience itself. As such Gautama's truth of pain does not admit of any objective test. We can, of course, view the pattern of human consciousness in a historical perspective; observe the expression which it has found at its highest level. The method is no doubt arbitrary to a degree; but no other approach seems practicable. And one fact emerges abundantly clear from an empirical retrospect of this nature: the fact that, at its most significant, human experience has so far found the tragic expression to be the most adequate representation of its reality. There are the bitter lamentations of a man of the land of Uz, whose name was Job: he feared God, and eschewed evil, and yet had to suffer torments of hell. There are the biblical Wisdom Books reminding us that human existence is a sore travail. There is Solon—a man-of-the-world and a sagacious law-giver—courting the disfavour of Croesus, king of Lydia, by observing 'consider the end of everything, and count no man happy till he is dead.' There is the 'Prince of Glory' kneeling in Gethsemane with his 'soul exceeding sorrowful unto death.' There is Kabir: he lived in God as 'fish live in water,' he was never haunted by that excruciating sense of exile from the Deity which drove so balanced a person as Empedocles to the verge of madness—and yet even Kabir lived long enough to admit that suffering is the common lot of men, believers and non-believers alike. There is Hamlet helpless against a 'sea of troubles'; Macbeth bearing the corpse of sleep 'to the last syllable of recorded time'; Lear 'a poor old man, as full of grief as age'—in tears; and Othello being goaded by a nameless and inscrutable cause into the murder of the thing he loved. There is Baudelaire. There is Dostoievsky . . . But one can prolong the tragic testimony indefinitely: it is overwhelming. The tragic, said D. H. Lawrence, is the most holding. It is the most holding, perhaps, because it is the most universal.

Universal, but not final. The truth of pain represents the point of departure for Gautama's critique of human experience, but it does not constitute the whole. Every truth has its limits: it is limited in its purpose, as also in its sphere of application. Gautama did not regard the truth of pain as the whole truth about life. He clearly defined its sphere, its purpose. 'So long, Monks,' he said, 'as I did not comprehend . . . the misery in the world as such, so long did I not discern the meaning of being enlightened . . . But, monks, when I fully comprehended . . . the misery in the world as such . . . then did I discern the meaning of being enlightened.' The understanding of 'the misery in the world as such,' is the primary condition of enlightenment. For Socrates the recognition of one's ignorance was the beginning of knowledge; for Gautama, on the other hand, the recognition of one's misery was the beginning of knowledge. But the beginning of knowledge is not the end of knowledge. To comprehend the fact of suffering in its universal implications is to discern the meaning of being enlightened; but the enlightenment itself, which is demonstrably a residual condition, is something-other-than-suffering. The recognition of the tragic motive as being the dominant motive in human psychology leads to an awareness which does not partake of the nature of the process through which it is reached: the ultimate issue of tragedy is something-other-than-tragedy. Some such baffling paradox was vaguely at the root of the classical concept of 'catharsis' as being the purpose of tragedy: such, too, is the paradox underlying Gautama's critique of life.

It is essential to understand this paradox, because failure to grasp it has led to many misinterpretations of Gautama's doctrine. The ignorant charge of morbidity, for instance, which is so often levelled against Gautama, may be directly attributed to a misconception prevailing with regard to his attitude to the problem of suffering. Gautama never claimed that the truth of pain is all-comprehensive; it is only the first of a series of four 'noble' truths he postulated; and he admitted that there are truths more comprehensive than the truth of pain—truths which include and transcend it. For pain itself is a condition, not a finality; a universal and poignantly significant condition, but still a condition. And to every condition, according to Gautama, there is a cause. The world we live in is a world of dependent origination; everything that exists in this world is bound in a chain of causal relationship; and all phenomena proceed from a cause. An infant crying in the night has cause for his complaint. The general phenomenon of 'the misery in the world as such,' also proceeds from a general cause. What is that cause, and is it possible to discover it and understand it 'as it really is?'

Gautama answered the question in the affirmative. We may, or may not, agree with his answer. Perhaps it tends to over-simplify the issue; but, for all its shortcomings, it remains a remarkable explanation—remarkable for its originality no less than for its depth and comprehensiveness. It anticipates—indeed, in some ways, it strikes a note that goes beyond—the modern psychological hypotheses. For

Gautama seems to have resolved the problem of human suffering in psychological terms, though it is necessary to add that his analysis is nearer Shakespeare than Freud. His diagnosis does not attempt to postulate a metaphysical and abstract cause for human suffering. It does not attempt to drag in the Deity and appeal to an arbitrary dualism of good and evil after the fashion of theologians. The tormenting sense of sin, which plays so important a part in the pathological philosophy of the Church Fathers and their latter-day devotees, does not enter into it: we do not hear of '*le spectacle ennuyeux de l'immortel péché*' as in Baudelaire, or of the 'doctrine of original sin' which is such a 'real and tremendous thing' to Mr T. S. Eliot. The critique is kept on a humanistic level: the explanation of human suffering is found in an immediate and intelligible cause. 'By oneself one suffers,' says the Dhammapada: by oneself—not because of Adam's fall from Grace, not because of the sins of our fathers. By oneself one suffers—but why? Because there is a contradiction in one's experience, a contradiction in one's approach to experience In contemporary phraseology we should describe it as a 'complex,' but contradiction is a more expressive and inclusive term. What is this contradiction which cuts across the whole of one's experience and transforms life into an unceasing agony?

The Fire Sermon hints at an explanation. The title of the sermon itself furnishes a clue to the issue: 'the Conflagration of the Senses' is precisely what constitutes a constant source of irritation and inquietude in the human psyche. We suffer because our hearts, our eyes, and, indeed, all other vehicles of our sense-perceptions are of the nature of a flame. We suffer because 'O monks, the knowledge . . . the feeling arising from contact with the visible . . . the audible . . . the palpable . . . be it pleasure, be it pain, be it neither pleasure nor pain, this also is in flames.' But how is the conflagration of the senses kindled? By what fire? By the fire of desire, says the Fire Sermon, above all by the fire of desire. The world we live in is a world of suffering because we approach it through the mediacy of desire. And desire is hunger; desire is deficit. And it is a hunger and deficit of a specific kind: it is a hunger for which there can be no satisfaction, a deficit which can in no way be filled. Desire reaches out towards the unattainable: it is, in fact, *soif de l'impossible*.

But why? Why does desire assume an insatiable hunger, an insatiable deficit, and an unquenchable thirst? We have the testimony of Nietzsche—a testimony all the more convincing because it comes from a man who, in his writings, appears to take a position almost diametrically opposed to Gautama. All desire, he says poignantly, yearns for eternity—yearns for deep, deep eternity. Desire, in other words, is not just necessity (what is necessary is, in fact, seldom desirable). Its demands are infinitely more exacting. To desire is inevitably to crave for eternity—for deep, deep eternity. And such a craving is foredoomed to frustration. For the world we live in is a world of transient things—in spite of Plato, in spite of Sankra. Every object which desire can 'grasp,' would like to 'grasp,' contains within it the inherent necessity of decay and dissolution. The 'grasping' mind itself is a component thing and, like the baseless fabric of Prospero's vision, dissolves, leaving not a rack behind. The *Brahman* of the Vedantic day-dream is no more eternal than 'the cloud-capp'd towers, the gorgeous palaces, the solemn temples, the great globe itself . . .' The immanent soul of the Platonic fancy is no more immortal than the face that launched a thousand ships. These fantasies are the desperate, touching, and hopelessly inadequate refuges built by the 'grasping' mind to escape the inherent necessity of dissolution.

Such is the 'noble' truth of the cause of pain.

But even this is not the whole truth. Gautama's critique goes further, postulating the cessation of pain. This may, or may not, satisfy us; but it is well worth some serious consideration. If the misery in the world as such, it argues, proceeds from a cause—from an immediate and intelligible cause—then it is perhaps possible, at least it is conceivable, that the cause can be extirpated. If we know and realize what is the source of irritation in our psyche, then we should try to remove that source of irritation. If we know and realize that we suffer because we approach life through the mediacy of desire, then the only thing to do is to learn not to approach it through the mediacy of desire. There are other ways of approach to life, other modes of experiencing reality, than desire. For desire itself, according to Gautama's analysis, is not a finality. He never posited it—as Schopenhauer posited his Will—as the Thing-In-Itself. Of the Thing-In-Itself he never pretended to have any definitive knowledge. Indeed, he appears to have believed that the knowledge of such would not be particularly useful even if it were possible; that the only kind of knowledge which is at once relevant to human purpose, and within the compass of human understanding, is the knowledge of relations, of conditions. As such, he represented desire as yet another condition, not the Thing-In-Itself; a deep-rooted condition which, as Nietzsche beautifully put it, 'is deeper yet than heartbreak'—but still a condition. And because it is a condition and not the Thing-In-Itself, it can be changed, it can cease to be. The change, the cessation, is not only possible; it is even inevitable. Inevitable that the grasping approach should change; inevitable, too, that desire should cease to be. For though it is true that desire 'is deeper yet than heartbreak,' there is also such a thing as a gradual wearying of desire. If the heartbreaks are persistent—and in the very nature of things, they must be persistent—then there is bound to come a moment in the curve of desire when it wearies of itself, ceases to be, through a process of self-exhaustion; a moment, in other words, when the 'grasping' mind has been so deeply annihilated by its successive failures that it automatically abandons its tendency to grasp. The infant crying in the night would, in the absence of the mother, cry itself to a profound silence; even Nietzsche could not pursue the Dionysian ecstasy of desire—desire to 'create something beyond himself'—further than a stupefying and speechless madness.

But this is not the kind of renunciation that Gautama contemplated. His is an 'Apollonian' rather than 'Dionysian' approach: it demands a conscious and intelligent renunciation of the element of irritation in the human psyche, a conscious and intelligent abandonment of the grasping attitude of mind. It is a renunciation which one has to achieve through one's own strenuous effort: by oneself one suffers, and by oneself one ceases to suffer.

Such is the 'noble' truth of the cessation of pain; leading us to the problem of the Way.

II

THE WAY

> There is a middle path, O Bikhus, avoiding the two extremes, discovered by the Tathagata—a path which opens the eyes, and bestows understanding, which leads to the higher wisdom, to full enlightenment, to Nirvana!
>
> *The Foundation of the Kingdom of Righteousness*

THE Way is of the world; and yet it is not a worldly way. The Way demands renunciation of a very ultimate nature; and yet the renunciation it demands is but an acceptance of the order of actuality. The Way sets forth limits to human experience; and yet these limits, when properly realized, become the vehicle of liberation for the consciousness. The Way is not the way of ardent love and desire; and yet it releases a tenderness which is beyond all measure. The Way leads through a wilderness of doubt and despair; and yet the 'wayfarer' in the very process of 'wayfaring' attains to a serenity free from all doubt and despair. The knowledge of The Way is born from the recognition of pain; and yet The Way transcends the condition of pain from which it is born.

The Way is a moral way; and yet it has no use for mere moralities. The Way is based on a concept of normality; and yet the 'Norm' it offers is not a measuring-rod, not just a social convention. The Way involves the working out of a technique of living; and yet the technique it contemplates is not a matter of ceremonious formalities, of codes of outward behaviour, of cultivating masks for presentation in public places. The Way is a way of detachment; and yet it is not the way of egotistic, self-centred indifference, which is merely an intellectual form of self-indulgence. The Way is simple, manifestly reasonable, and within the grasp of the humblest of men; and yet subtle, hard to perceive, hard to know—and harder still to follow.

The Way is a paradox.

The Way is of the world because it concerns itself with the most universal, the most immediate problem of life: the problem of grief and tribulation, of ill, sorrow and distraction. It is of the world because the calm, the insight, the enlightenment and Nirvana to which it tends are not esoteric states, and are relevant only in the here and now. And yet it is not a worldly way; and the worldly-wise, both those who seek their land of heart's desire in the here and now, and those who project it into an imaginary hereafter, have always found it unsatisfactory. It is not a worldly way because it does not concern itself with the mere acquisition of sensations, physical or spiritual, positive or negative, of this world or of the next; because it avoids the two extremes which have a basic identity of purpose—'that conjoined with the passions, low, vulgar, common, ignoble, and useless, and that conjoined with self-torture, painful, ignoble, and useless.'

The Way demands renunciation, a difficult and irredeemable psychological surrender: the surrender of one of the most deep-rooted habits of the human ego—the habit of grasping. Other religions, other faiths and philosophies have also demanded such a surrender. There is, however, a qualitative difference between their demands and the demand which is implicit in Gautama's conception of The Way. The latter does not make the demand for renunciation conditional upon any arbitrary ethical principles or compensatory promises. It asks for the abandonment of the habit of grasping because it recognizes the contradiction inherent in the acquisitive approach; recognizes the illusory nature of the human grasp—the illusory nature, indeed, of the very ego that wants to grasp; recognizes, in the ultimate analysis, the utter impossibility of grasping anything at all in a universe of sense and succession. It asks for the abandonment of the habit of grasping because it knows 'that it is in the very nature of all things most near and dear to us that we must divide ourselves from them, leave them, sever ourselves from them.' The renunciation which is demanded is thus no more than an objective acceptance of the actual conditions of existence.

These conditions necessarily impose limits on human experience. The periphery of human experience—both in its actuality and its potentiality—is strictly finite, and as such, limited. It is limited in time; it is limited in space. And although these limits are not stationary, but always changing, yet the measure of that change is always finite and subject to limitations. The Way, therefore, admits the essentially limited character of our experience. It goes further: it suggests that the only manner of expanding the horizon of our awareness is by accepting its finiteness, and the finiteness of our vision. Do not look long, it says, and do not look short; in other words, avoid long-sightedness as well as myopia. For The Way can be perceived only by a normal vision; and the normal vision is a finite vision. It is neither macrocosmic nor microcosmic. It does not concern itself with the infinite or the infinitesimal. Its principal focus is the finite universe—that strangest of all paradoxes which is at once a part of the infinite, and contains an infinity within it. Do not look long, and do not look short. For between the tip of one's nose and the darknesses which stretch beyond the remotest star there are a multitude of finite objects and distances, which our eyes are often apt to overlook, yet which it is imperative not to overlook. It is only by restricting our focus to a finite universe that we can hope to attain to some measure of lucidity and freedom of vision. Freedom, said Engels, is the recognition of necessity. Gautama would have agreed, though he would probably have added that necessity can be apprehended only by a vision that is neither long-sighted nor myopic. Such is the essence of a truly human perspective—it is the essence of The Way also.

The Way is weary of ardent love and desire. And why? Because love and desire cannot be adjusted to necessity, to the conditions of a finite universe. They demand too much or too little; look too far or too near. And why? Because they are, in a more than mere symbolic sense, blind. And blindness separates, creates insuperable barriers. Ardent love and desire separate one from the very object which one so loves and desires. The Way is weary of them also because they are still involved in personality, in the ego: they represent, in fact, one of the most subtle attempts on the part of the ego to save itself—by losing itself in something-other-than itself. And yet, of course, the ego can never really lose itself in the object: it merely subjectifies and appropriates it. Ardent love and desire are insatiably acquisitive, and because they are thus, they stultify that all-pervasive tenderness which is not born of hunger, but of understanding; which, because it is not blind, does not create barriers, but unites. To abandon the way of love and desire, therefore, is to remove barriers, to release a tenderness, which, because it is not involved in the ego's hankering for possession, and is free from the limitations of personality—

> *. . . Plane sur la vie, et comprend sans effort*
> *Le langage des fleurs et des choses muettes!*

However, to arrive at such an all-comprehensive understanding, to realize in one's being such a quality of immediate and effortless sympathy, it is necessary to pass through a desolation of doubt and despair, to know the depths of human suffering in their utter nakedness. For, says Gautama, 'so long, O monks, as I did not comprehend, as it really is, the satisfaction in the world as such, the misery in the world as such, the escape therefrom (i.e. misery) as such, so long did I not discern the meaning of being enlightened . . . But, monks, when I fully comprehended, as it really is, the satisfaction in the world as such, the misery in the world as such, the escape therefrom as such, then did I discern the meaning of being enlightened in the world. Then did knowledge and insight arise in me . . .'

The Way is a moral way; and yet Gautama himself repeatedly discouraged those among his followers who wanted to interpret his teachings on a purely ethical level, divorce his precepts from their psychological background, and turn his critique of experience into a mere formula. In the *Brahmajala-sutta* he declares quite frankly that those who praise him on account of his morality praise him for the wrong reason. 'It is in respect only of trifling things,' he says, 'of matters of little value, of mere morality, that an ignorant man, when praising the Tathagata, would speak.' The Eightfold Path of right views, right intention, right speech, right action, right livelihood, right effort, right mindfulness, right concentration, has, therefore, little meaning when considered in isolation from his psychological interpretation of the nature of human experience. It is only when we correlate it with the whole analytical process of his thought that it acquires a vital, highly original significance. For instance, the phrase 'right views,' which forms the basis of the Eightfold Path, is a mere ethical cliché considered by itself. But it ceases to be an ethical cliché when it is realized that Gautama did not use it in its ordinary didactic implication; that by 'right views' he seems to have implied a perspective of life free from such common 'delusions' as belief in an omnipotent Deity, in the permanence of the human ego, in personal immortality and posthumous compensations—a perspective, in other words, which is not distorted by one's more obvious desires and wishes. Thus The Way is moral only in a very novel sense: it is moral in the sense that, quite independent of any metaphysical and supernatural considerations, it places the responsibility of action upon the individual; that, taking psychological observations as its starting point, it finds 'by oneself is wrong done, by oneself one suffers. By oneself wrong is left undone; by oneself one is purified. Purity and impurity belong to oneself; no one can purify another.'

The Way reduces morality to a function of human intelligence, to a kind of honest self-criticism and censorship. And though the notion of 'right' underlying The Way presupposes a concept of normality, the Norm which is set forth is not just a measuring-rod, not a moral dogma. All that it implies is a concept of centrality in human experience. The Norm is not meant to be a formula nor a code of reference, but a focus—a point of integration for the psyche. Human awareness, one might say, evolves within the spiral of increasing perception; and the Norm is as the centre about which the spiral revolves in its ascent.

The Way obviously does involve a technique of living, an *ars vivendi*. Equally obvious is the fact that the technique of The Way is not a matter of outward form. Gautama, unlike Comte, never attempted to 'strengthen the feeble altruism of human nature by exalting Humanity as the object of ceremonial worship,' and did not spend his twilight years 'devising for this Religion of Humanity an intricate system of priesthood, sacraments, prayers, and discipline.' Life in the Order during his Ministry seems to have been remarkably free from ceremony and ritual, for such a quasi-religious body.

Yet The Way insists on the need for discipline; not a ritualistic and ceremonious discipline, but a discipline of the whole being. There is nothing esoteric about the discipline of The Way: for Gautama 'preached the truth without making any distinction between exoteric and esoteric doctrine.' The discipline only emphasizes 'right mindfulness and right concentration.' The purpose of this discipline is not, as in Yoga, the attainment of a subjective state akin to 'deep and dreamless sleep,' or a sense of unity with the Absolute; that peculiar interpretation of The Way, the concept of '*Dhayana*' for instance, is a latter-day innovation; it belongs to men like Bodhidharma who went to China and founded what developed into Zen Buddhism, a cult which for all practical purposes is indistinguishable from Yoga. Actually, both in its method and purpose The Way is so utterly different from the method and purpose of Yoga that it is difficult to understand how Louis de la Vallée Poussin could confuse the two, saying: '*Le Bouddhisme est une certaine forme de yoga ou ascétisme-mysticisme; il est né de la cristallisation rapide de donnée généralement Indienne sous l'influence de Câkya Muni*, "*le grand religieux.*" '

This confusion results, of course, from a complete misunderstanding of the issue. Yoga aims at achieving oneness with the Supreme, with the Immanent Brahma; and it proceeds by way of subjectification of the object. The Way, on the other hand, if it does not deny the reality of Brahma, at least expresses very grave doubts as to his importance; as such it is not very much concerned with chasing the Absolute. The method of The Way is also almost diametrically opposed to the method of Yoga: it tends to approach 'the world within' from as objective an angle as is humanly possible, to bring the chaotic tendencies of the psyche under some kind of a rational control by cultivating the habit of dispassionate introspection and meditation. The Way seeks tranquillity rather than exultation, the peace which resides in detachment rather than the ecstasy of imaginative identification, which constitutes the *élan* of the mystic.

The whole essence of The Way is expressed in one of Gautama's very last sermons which has been given elsewhere, but which may pertinently be repeated here. 'Herein, O mendicants, let a brother, as he dwells in the body, so regard the body that he, being strenuous, thoughtful, and mindful may, whilst in the world, overcome the grief which arises from bodily craving—while subject to sensations, let him continue so to regard the sensations that he, being strenuous, thoughtful, and mindful, may, whilst in the world, overcome the grief arising from the craving which follows our sensation—and so also as he thinks or reasons or feels let him overcome the grief which arises from the craving due to ideas, or reasoning, or feeling.'

The sermon doubtless sets forth an ideal of detachment. However, the detachment that is demanded should not be confused with the negative ideal of indifference. The distinction between the two ideals may be subtle, but it *is* there: it is to be found in Gautama's emphasis on the necessity of 'being mindful and thoughtful.' The mindfulness and thoughtfulness which are contemplated in The Way assume a high degree of sensitiveness to the world around one; they assume a comprehending awareness which is ready to admit the fullness of knowledge into itself. For, demonstrably, it is impossible to be constantly 'mindful and thoughtful' when one is indifferent. The detachment of The Way, then, is in reality, sensibility in its most crystallized and awakened form. Such a crystallization may, or may not, be practicable for humanity *en masse*. Still the fact that Gautama in his own person almost succeeded in realizing an awareness of this order, is something over which one may justifiably rejoice.

The Way is profoundly paradoxical in its realization as well as in its practice. Yet the paradox is one which is by no means beyond the grasp of human intelligence: to resolve it merely requires the use of what Descartes called '*le bon sens*' and described as being '*la chose du Monde mieux partagée.*' Only, of course, there is the good sense which is an active agency, and the good sense which lies dormant and passive as an unrealized potentiality. Till now, at any rate, active good sense has been conspicuous mostly by its absence from human affairs. That is why The Way is hard to perceive, hard to know—and harder still to follow.

There is always a world of difference between the mere recognition of The Way and its attainment.

III

THE VOID

We are such stuff
As dreams are made on, and our little life
Is rounded with a sleep . . .

Prospero in *The Tempest*

MAN, Schopenhauer was fond of saying, is a metaphysical animal. Schopenhauer, it is well known, was an ardent admirer of Indian philosophy; he had drunk rather deeply of its sad wisdom; and probably he was thinking of the Hindus in particular when he pronounced this acute judgement on the nature of man. For no other people under the sun have been so addicted to metaphysics as the Hindus; it is their hereditary passion; and there is scarcely anything in their complex culture which is not permeated with it. The Hindu is, in truth, the metaphysical animal *par excellence*. This is not to suggest that he is fundamentally different from the rest of mankind; or even that, because of his metaphysical preoccupations, he in any way neglects the physical side of life. Far from it! He takes the physical ritual of existence as seriously as anybody else; and indeed, if anything, he attends to the mundane processes of eating and evacuating, breathing and procreating with an even greater relish and gusto than an ordinary *homme moyen sensuel*. However, where he differs from the latter is in the peculiar habit he has of giving a transcendental twist to his physical functions.

This transcendentalism may partly be attributed to the quite extraordinary, and often exasperating, vanity of the Hindu mind: a vanity which, at root, is a subtle form of megalomania. It enables the Hindu to surround himself with a cosmic aura, and to invest the most trivial details of his life with a sense of importance for which there would appear to be not the slightest justification. Moreover, metaphysics serves him as a kind of intoxicant: an intoxicant in some ways even more stimulating than the potent essence of Soma, whose magical qualities have been celebrated with such extravagant lyrical abandon in the Sama Veda —more stimulating since it dispenses with the mediacy of the senses and directly attacks the brain. His transcendentalism renders the Hindu almost completely immune from that most universal of human afflictions—*ennui*.

The Hindus of Gautama's days were even more strongly addicted to metaphysics than the Hindus of to-day. Gautama was one of the very few thinkers of his age who successfully resisted the temptation of abstract beatitudes. It must have required no small amount of courage and strength. In fact, he did more than merely resist the temptation: he actually raised a voice of protest against this spiritual form of indulgence, and pleaded for temperance. For all practical purposes it was a voice in the wilderness, since even those who cared to listen to him did not take the trouble to understand him. Yet the fact that it was raised is in itself pregnant with meaning.

Gautama was not at all a metaphysician. Indeed, he never concealed the fact that he regarded metaphysics as an irrelevance—an intellectual luxury which might be permissible in some happier world, but not in this best of all possible worlds, where the tears shed by all sentient creatures exceed 'the water in the four oceans.' He described the endless wrangles of different schools of metaphysicians as 'the thicket of theorizing, the wilderness of theorizing, the tangle, the bondage and shackles of theorizing.' For himself he had no use whatever for theorizing. His method of attack is always direct and concrete; his arguments are lucid and lively, relieved, as they often are, by a gentle irony which some of his contemporaries considered as erring on the side of levity. In an age and a country given to excessive indulgence in abstractions and day-dreams, he insisted on the too, too solid world of pain; insisted on the necessity of facing the urgent problem of conduct; insisted on placing first things first. This was his distinctive contribution to the cause of humanism.

Obviously, then, it would be futile to look for anything like a completed system of metaphysics in his critique. The bulk of the Buddhist metaphysics, such as we know it today, is the product of the fertile imagination of the Mahayana School, or the Greater Vehicle, just as the formalized morality and ethics is mostly the creation of the puritanism of the Hinayana School, or the Lesser Vehicle. Gautama's own attitude to metaphysics was akin to that of Hume and his followers: he mistrusted speculative philosophy because he considered it to be based on insufficient data, to be no more than an infantile pastime of building castles in the air. This does not justify the inference—often too easily drawn by his hostile critics—that his reluctance to commit himself with regard to the metaphysical issue was due to his desire to avoid the mental effort necessary to resolve it, and that his baffling silences were nothing but a confession of his lack of a comprehensive and coherent world-view. In point of fact, during the years between his Renunciation and his Enlightenment Gautama had followed the metaphysical method as far as the most advanced among the metaphysicians of his age—only to realize its hopeless limitations. And he was honest enough to confess them. He had rejected the metaphysical approach for the same reason for which he rejected mysticism: because he did not think it touched the heart of the human problem. His critique outlines a world-view, a remarkably coherent and comprehensive world-view; only it is not a metaphysical world-view. In the last analysis perhaps, metaphysics is an intellectual equivalent of the Myth. And the problem before man, as Gautama conceived of it, was not merely to rationalize the Myth, or find an intellectual substitute for it, but to dispense with its mediacy altogether—to outgrow the Myth-Complex.

Even to refute metaphysics, however, one has to be metaphysical. Nobody disliked abstruse ratiocination more than Gautama; and yet he could not help having endless arguments about precisely those matters which he regarded as 'irrelevant.' In these arguments, naturally, he had to adopt a somewhat-negative position: one cannot be expected to be positive about things which one considers as of little importance.

There was the problem of God, for instance. It was in every way a tantalizing problem. To begin with, it was a difficult problem to define. Different people interpreted the term in different ways. A most bewildering diversity of views prevailed about God; beliefs ranging from polytheism to atheism, from a rigid and refined monism to the most primitive animism. In their attempts to reach a conclusive and all-embracing definition of the issue many a brilliant intellect had been driven to the verge of insanity. And yet no satisfactory definition had been realized. In the very nature of things, it could not be realized.

Gautama's attitude to God was simple and straightforward. Of God in the theological sense—that is in the sense of the First Cause, *causa causans*—he frankly admitted he knew nothing. He went even further, saying that he did not particularly want to know who was the ultimate author of this universe, and how and why it had originally come into being. He did not want to know because he did not think it either possible or important to know. He was content to take the universe for granted, to make the best of it as it was, and to understand its internal laws and relations without troubling about the ultimate 'how' and 'why.' It was not a very heroic attitude, but it was the only attitude compatible with common sense. As he said to one of his disciples who had decided to leave because the Tathagata would not reveal 'the beginning of things': 'Whether the beginning of things be revealed or not, the object for which I teach the Norm is this: that it leads to the thorough destruction of ill for the doer thereof.'

Of God in the sense of the personal Creator, the Deity ruling over the universe, the Dispenser of the moral law of retribution and reward, Gautama was more than sceptical. There is scarcely anything in his more authentic dialogues which can be suspected of implying belief in a personal God either in the theological or mystical interpretation of the term. On the other hand, there are many categorical statements to show that he found it impossible to believe in a personal Deity as a matter of faith, and that as a matter of experience it never became real to him. There is the evidence of the *Tevigga Sutta*, for example. 'But then, Vasettha, is there a single one of the Brahmans versed in the Three Vedas who has ever seen Brahma face to face?' Gautama asks a young Brahman. On the latter's replying in the negative, he remarks: 'So that the Brahmans versed in the Three Vedas have forsooth said thus, "What we know not, what we have not seen, to a state of union with *that* we can show the way, and can say: 'this is the straight path, this is the direct way which leads him, who acts according to it, into a state of union with Brahma.' " Now what think you Vasettha? Does it not follow, this being so, that the talk of the Brahmans, versed though they be in the Three Vedas, is foolish talk?'

Further in the same discourse he says: 'Just, Vasettha, as though a man should make a staircase in the place where four roads cross, to mount up into a mansion. And people should say to him, "Well, good friend, this mansion to mount up into which you are making the staircase, do you know whether it is in the east, or in the south, or in the west, or in the north? Whether it is high or low or of medium size?" And when so asked he should answer "No." And people should say to him, "But then, good friend, you are making a staircase to mount up into something—taking it for a mansion—which all the while you know not, neither have seen." And when so asked he should answer "Yes." Now what think you, Vasettha? Would it not turn out, this being so, that the talk of that man was foolish talk?' The Brahmans are then compared to the man who constructs 'a staircase to mount up into something—taking it for a mansion—which, all the while, he knows not, neither has seen,' and Brahma to the mansion the existence of which is highly problematical.

But this is not all. Gautama's agnosticism cuts much deeper. There was still one way open to him whereby he could have retained God in his scheme of things—that is, on the ground of expediency and necessity. The eighteenth-century Deists like Voltaire maintained that if there were no God it was necessary to invent Him; they evidently believed this, because otherwise it is impossible to establish any form of moral order in the world. Similarly, modern pragmatists like William James and John Dewey have stood by the Deity because they have found that the concept 'works.' Gautama's position on this issue is in striking contrast to the utilitarian Deists. He did not regard God as a necessity. Gautama regarded the universe-in-manifestation as a self-sufficient actuality that could, and did, function without divine intervention. This position was already implicit in Kapila's dualistic philosophy. But while Kapila had been anxious not to cause offence to the hierarchical authority, and had left the whole matter somewhat equivocal, Gautama restated the problem with absolute clarity. He considered it perfectly possible for man to evolve a scheme of moral sanctions, to develop a sense of individual responsibility, without any reference to an external and omnipotent authority. The purpose and function of human intelligence then, is not so much to attempt the impossible task of explaining the ultimate origin of the universe, as to observe and comprehend the order of its manifestations 'as it really is.' This is all that Gautama himself appears to have attempted.

Very closely linked with the question of God, was the question of soul—a permanent and immortal soul. It was in Gautama's days—as it has been ever since—a burning question. Also a difficult question to discuss, since it directly involves human wishes and *amour propre*. And wherever human wishes and *amour propre* are at stake, it is idle to expect any objectivity. Gautama seems to have been one of the very few men who succeeded in discussing the problem with a certain measure of dispassionateness. And because he discussed the problem dispassionately, his analysis once again led to a negation. He denied the existence of a permanent soul, or self, or Atman as the Brahmanical metaphysicians described their fanciful creation. In the *Anatta-lakkhana-sutta*, the second sermon he preached to the five monks in the Deer-park of Isipatana at Benares, he explicitly declared that there is nothing in all the elements which constitute an individual, that corresponds to the notion of an immanent and immortal soul as conceived by the Vedantists. 'The body, feeling, perception, consciousness, and the aggregates,' he insisted 'are all soulless.' The only sense in which he was prepared to admit the idea of the soul, or the self, was in its empirical sense; in the sense in which the enigmatic Buddhist sage Nagasena was to define it in the course of his famous argument with king Milinda (Greek Menander) of Bactria—that is, as a component thing which is not at all of the nature of a permanent and unchanging reality underlying our experience, but merely represents the synthesis of the totality of our sensations, perceptions, intuitions, bodily and mental reactions at any given moment at any given place. And because the soul of human experience was for him a component thing, he held that it was not an absolute and static reality, but a sequence of kaleidoscopic patterns. In other words, he regarded the self, or the soul, as a mirror; a mirror of a special and unusual kind, a mirror that does not exist apart and independently of what it reflects, but comes into being and exists only by virtue of what it reflects. Thus, on the one hand, Gautama's conception of the soul, or the self, was nearer Heraclitus the Obscure than the Vedantic revivalists like Sankra; on the other, it corresponded more closely to the views held by Democritus and his sadly neglected master Leucippus, than Plato.

This peculiar conception of the soul is so unequivocally expressed by Gautama that there is hardly any room for misunderstanding on the crucial issue. Rhys Davids observes: 'The position is so absolute, so often insisted on, so fundamental to the right understanding of primitive Buddhism, that it is essential there should be no mistake about it. Yet the position is also so original, so fundamentally opposed to what is usually regarded as religious belief, both in India and elsewhere, that there is a great temptation to attempt to find a loophole through which at least a covert or esoteric belief in the soul, and in the future life (that is of course, of the soul) can be recognized, in some sort of way, as part of so widely accepted a religious system. There is no loophole, and the efforts to find one have always met with unswerving opposition, both in the Pitakas themselves and in extra-canonical works.'

In view of Gautama's categorical denial of the permanent nature of the soul, it is difficult to understand how anybody could justifiably credit him with belief in transmigration. Yet from the time of the earliest compilers of the Canon to the present day, it has been the custom to attribute this absurdity to the Tathagata. Even Rhys Davids, after commenting that 'it would not be possible in a more complete and categorical manner to deny that there is any soul—any entity, of any kind, which continues to exist, in any manner, after death,' than was done by Gautama in the *Brahmajala-sutta*, goes on somewhat surprisingly to argue that 'Gautama had not been able to give up the belief in transmigration.' If Gautama had not been able to give up the belief in transmigration, then one can only conclude that he was either so muddle-headed that he did not know what he believed, or was being deliberately dishonest. For the dogma of transmigration in its Brahmanical form at least has the saving grace of being a consistent, and up to a point, satisfying, doctrine; but in the form in which it has been retained to support the Buddhist eschatology, it is the most obvious piece of chicanery. Belief in transmigration when combined with belief in a permanent soul still remains very much of a wish-projection and an illusion; but at least it is a logical and consistent illusion. Belief in transmigration without a corresponding belief in an immortal soul is an illusion which is at once illogical and inconsistent.

In all other respects Gautama's world-view is so transparently sound and reasonable, that it is hard to believe, in this particular case, he should suddenly have abandoned his rational outlook, thus annihilating the whole purpose of his philosophy. On the other hand, it seems much more credible, and is indeed highly probable, that the references to the doctrine of transmigration which are to be found in the Buddhist Canon—and they are legion—have been introduced there by the compilers. In the first instance this strange doctrine, which figures in all subsequent developments of Buddhism as a religion, must have been incorporated into the original teaching of Gautama as a concession to, and compromise with, the prevailing popular belief in rebirth, the idea being to render Buddhism acceptable to the masses. Moreover, the Buddhist law-givers must have perceived that the doctrine in question, however arbitrary it might be, was an excellent device for strengthening their hands in the task of enforcing 'the Good Law' among the faithful. For, demonstrably, the doctrine that, 'as soon as a sentient being (man, animal, or angel) dies, a new being is produced in a more or less painful and material state of existence, according to the "karma," the desert or merit, of the being who had died,' is calculated to exercise a restraining influence over all refractory and rebellious spirits in the flock, as well as to promote social stability. Indeed, in some ways this doctrine is far more effective in realizing its moral ends and making the people follow the straight and narrow path than the doctrine of fire and brimstone, and the thundering voice of that 'righteous judge,' Jehovah, who, simply because 'He is angry every day,' ceases after a time to strike terror into the souls of all but the most timid.

Finally, it is possible that the concept of survival and rebirth was foisted upon Gautama's doctrine by his interpreters, who were anxious to discover some tangible explanation for his 'ethical insistence.' Even to-day highly advanced and idealistic writers—men like Dr Radhakrishnan—seem to experience an insuperable difficulty in conceiving of a moral philosophy devoid of all compensatory motives. It is, therefore, easy to imagine the difficulty experienced in this matter by interpreters of an earlier epoch. For all their good intentions, they could not understand for their own sake the practice of such human qualities as goodness, truth, virtue, and compassion—could not understand, as Spinoza puts it, 'that blessedness is not the reward of virtue, but virtue itself.' Yet, as has already been stressed, it was precisely in this that the originality of Gautama's morality consisted: it was not dictated by any consideration of the hereafter, but by the compelling exigencies of existence in the here and now—a position which Aristotle also takes up in his Ethics. For Gautama, The Way was not something apart from the goal: The Way *was* the goal.

The Way was not contingent on a knowledge of the 'eternal order' immanent in the pattern of the universe: rather, it was contingent on a proper realization of the manifest aspects of the pattern. From a narrow philosophical point of view, this position might be interpreted as being tantamount to complete repudiation of philosophy as a mode of arriving at truth. Not unnaturally, certain strict philosophers have been inclined to see in Gautama an early avatar of Sophists like Gorgias and Hippias. To do so, however, is to misrepresent and misunderstand Gautama. His position was fundamentally different from that of the Sophists—in India they were known as 'Eel-wrigglers'—with whom he had little sympathy. His critique of philosophy repudiates a particular approach to philosophy, not philosophy as such. The difference is important. Indeed, his very negations are pregnant with a definite and positive philosophic purpose. There is his negation of the Vedantic Absolutism, for instance. At the time when it came, it served as a positive and liberating influence. It brought a breath of fresh air into the exhausted philosophic atmosphere of India. For ever since Vyasa fell to the lures of Absolutism and landed in the labyrinthine cul-de-sac of Uttara Mimansa, Indian thought had not been able to find its way back out into the open air. The bulk of Indian thinkers had got so perfectly reconciled to their cul-de-sac after a time, that they regarded any attempt to get away from it as, if not actual heresy, at least something in thoroughly bad taste. Their prolonged preoccupation with what George Santayana aptly calls 'the musty Absolute' had reduced Indian philosophy to the most abject state of anaemia. Gautama's departure from Absolutism was, therefore, highly opportune: it had the effect of galvanizing Indian thought into new life.

Equally positive was Gautama's affirmation of the principle of causality, or the doctrine of *Paticcasamupada*—the doctrine, that is to say, 'that all *dhamma* [phenomena, mental and physical] are *Paticcasamppana* [happen by way of cause].' This was not an entirely new doctrine. The Chain of Causation is mentioned both in the Yoga and in Samkhya systems. However, it is only in the Great Discourse of Causation (*Mahanidana-sutta*) of *Digha-nikaya* that it first emerges with absolute clarity as the basis of a world-view. 'It is only in the Buddhist Nikaya,' says Rhys Davids, 'that we come up against the actual effort itself of the human mind to get at a more scientific view of the world-order—an effort which is marked with the freshness and vigour of a new fetch of intellectual expansion, and the importance and gravity of which is affirmed with the utmost emphasis, both in the earliest records and in the orthodox literature of ten centuries later.' Further, the principle of causality in Gautama's world-view is allied with a still more important principle—the principle of flux, of sense and succession, of change and Becoming. His was one of the earliest formulations of a dynamic view of life. He unfolded before the human mind the vision of vast and fluid horizons—the vision of a universe literally without beginning or end. It was like awakening to a new and apparently endless dawn.

In formulating his dynamic view of life, however, Gautama did not lose his sober judgement. He seems to have been careful to avoid the pitfalls to which most dynamic philosophies are liable. He never abandoned himself to that sense of irresponsibility which led Heraclitus—in many other respects a wise, and even profound man—to interpret his dialectics of universal flux as a philosophic justification for the ideal of perpetual war and strife, conflict and unrest. On the contrary, paradoxical though it may sound, the fact that we live in a world of flux and movement was, for Gautama, all the more reason for cultivating precisely those habits of mind which the 'Obscure' sage of Ephesus so heartily despised—that is to say, tranquillity, peace, contemplative detachment, stillness and poise. Unlike Bergson, Gautama was never carried away by the inebriating exuberance of his own phraseology: he was far too deeply aware of the inordinate squalor and misery, the unavailing suffering and waste, the unutterable frustrations and shipwrecks of life to offer mere verbal palliatives. His was a compassionate view of life.

'All that we are is the result of what we have thought: it is founded on our thoughts, it is made up of our thoughts'—reads a passage in the Dhammapada. It might almost be interpreted as the idealism which is implicit in the Cartesian proposition of *cogito ergo sum*. But Descartes's proposition is open to one rather facile, but all the same very pertinent, criticism—the criticism which Mr Aldous Huxley voices through one of the characters in *Eyeless in Gaza*. '*Cogito ergo sum*,' ruminates Anthony Beavis (ruminates, let it be added, over a typewriter—apparently for the convenience of posterity), 'But why not *caco ergo sum? Eructo ergo sum?* Or, escaping solipsism, why not *futuo ergo sumus?* Ribald questions. But what is "personality"? ' And after a long and discursive meditation (duly reduced to typescript), he decides that the most valid proof that we exist is that our bodies are *there*. 'In the swamp and welter of this uncertainty,' he concludes with evident satisfaction at having solved the riddle of personality, 'the body stands firm like a Rock of Ages.

Jesu, pro me perforatus,
Condar intra tuum latus.

Even faith hankers for warm caverns of perforated flesh. How much more wildly urgent must be the demands of a scepticism that has ceased to believe even in its own personality! *Condar intra* MEUM *latus!* It is the only place of refuge left to us.'

Condar intra MEUM *latus!* Surprisingly, this also happens to be the refrain of one of Gautama's dialogues. Surprisingly—because usually one does not associate such views with the Tathagata whose very glance, M. René Grousset tells us, 'seems to be plunged back into the interior Essence, into the ineffable Buddha-state—a glance which sees into the beyond, a glance that haunts, heavy with all the metaphysical thought of the Mahayana, weighted, in its ardent fixity, with all the virtualities of the cosmic play, and vanishing within into the vacuity of substance.' Indeed, after reading such ecstatic eloquence, one would hardly dare to attribute the dialogue in question to the Buddha, were it not that irreproachable authorities have already done so. For in this discourse, Gautama not only represents the body as 'a Rock of Ages,' but he suggests that it contains within its narrow confines the whole content of this vast universe of sense and succession—its myriad stars, and its empty spaces, and its unfathomable abysses. 'Verily, I declare unto you,' he says, as though forestalling the objections of some unborn Anthony Beavis, 'that with this very body, mortal as it is, and only a fathom high, but conscious and endowed with mind, is the world, and the waxing thereof and the waning thereof and the way that leads to the passing away thereof.' If this is an idealistic position, then Protagoras and Hume must have been idealists.

Gnothi seauton! Socrates declaims the inscription from the oracle at Delphi for the general enlightenment and edification of mankind. The war about the origin, nature and validity of knowledge which Descartes started with his '*cogito ergo sum*' is still in progress, though mighty heroes like Leibnitz, Locke, Bishop Berkeley, Hume, and Kant fell and were buried long ago. Kant's *Critique of Pure Reason* destroyed all that was left intact of the philosophical Troy; and Helen must now be sought within, not without us. Fichte seriously informs us that 'the world is his idea.' Schopenhauer repeats Fichte's dictum with an even more emphatic shake of his head. But, we are tempted to ask, what does it all mean? And where does the philosophical flight lead to? What is the 'self' which we ought to know about and of which we should take care? What is the thing that cogitates? Where precisely must we look for the seat of Pure Reason? It is all a tremendous question-mark. And the philosophers seldom come to the aid of bewildered philistines. They evade the issue, or answer in terms which are far too abstruse and vague to be satisfying. It is safer to return to Gautama. With him we at least know where we stand. His answer is straightforward and intelligible. It is not an heroic answer. Some would probably regard it as bordering on frivolity—a weakness not at all becoming in a Tathagata, in a 'Knower of The Way.' But there is serious purpose behind Gautama's frivolity, if frivolity it is. His answer has the great advantage of being intelligible, whereas most other philosophers are obscure; the truth it contains is—in a certain measure—actually verifiable. *Condar intra* MEUM *latus*. The whole of man's knowledge and experience begins and ends in what Father Hopkins describes as 'his bone-house, mean house.' This is Kant, minus, fortunately, his 'Transcendental Aesthetic and Transcendental Logic.' At least we know what 'within us' means. There is no ambiguity in the term: Gautama's words are crystal-clear—'Verily, I declare unto you, that with this very body, mortal as it is, and only a fathom high, but conscious and endowed with mind, is the world, and the waxing thereof and the waning thereof, and the way that leads to the passing away thereof.'

Mrs Rhys Davids protests that this amounts to 'an irrational denial of the man as man,' it is reducing man 'to his instruments'; it is 'the rejection of divinity in the self, the self himself, the man, the person . . .' Perhaps it is all these things. But it also means making man the measure of his world.

'In reality,' declared Democritus, more than a hundred years after Gautama's death, 'there are only atoms and the void.' In the ultimate analysis, this is no doubt true. However, the ultimate analysis is so much of a simplification that it bears little recognizable likeness to reality as our eyes are capable of perceiving it. Democritus' ruthlessly analytical vision had failed to perceive a whole world of things. In a sense it had even failed in being properly analytical. It had failed to perceive that atoms are not atomic; that the mustard-seed is a microcosm. 'The atom,' Vasubandhu, a worthy exponent of the *Yogacara* system of Mahayana metaphysics, was to point out in the fifth century A.D., 'is not proved as such.' He meant, of course, that the atom is not an ultimate unit; that it is not one thing, but many; and that the whole concept of the atomists, in so far as it claims to be the concept of ultimates, is contradictory and absurd.

What then is there in reality?

LANDSCAPE

(attributed to Ma Kuei.
From 'Civilizations of the East' by René Grousset)

Gautama, who preferred not to look long and not to look short, answers that in reality there are only component things. No matter how far and how deep we carry our analysis of reality we still stand face to face with component things capable of still further analysis; no matter whether our eye travels outward, in ever-widening circles, or inward, in ever-narrowing circles, it can see only patterns which are made up of still other patterns. All our knowledge is a knowledge of relations, of conditions—not finalities. This is not a doctrine of *Maya*, or Illusion, as it is so often represented: it is simply a doctrine of relativity. For all relations are by their very nature relative—relative to the conditions which determine them. We are back again at the causal formula.

What are the elements which go to the making of these relations? What lies beyond and behind the conditions? These are tantalizing questions. The two chief schools of Buddhist metaphysics which grew up with the Greater Vehicle have each offered their own answers. There is the *Sunyata* doctrine of the Madhyamika School, founded by Nagarjuna, a metaphysician who lived in northern Deccan in the first century of our era. *Sunyata*, literally vacuity, is interpreted in the *Garland of Flowers*, a mystical work written in the second or third century A.D., as the Essential Nature of things, the *tathata* in Sanskrit. 'This Essential Nature,' observes M. Grousset, 'will constitute a kind of Divinity appearing in the very bosom of the Nagarjunan vacuity. It will present itself to feeling, if not from the metaphysical point of view, as an equivalent of the Absolute, or, if you like, as an Absolute no longer superior to phenomena, but entirely inherent in them, the Absolute as the actual processus of things. On the surface and in the bosom of the unfathomable vacuity, the bottomless ocean supposed by Nagarjuna, the phenomena are the ocean considered as waves; the Essential Nature is the waves considered as the ocean.' Later on, *Sunyata*, or the Essential Nature, became 'under the name of *Prajna Paramita* or Perfection of Sapience a kind of revelation of truth or, if you like, of the Buddhist holy wisdom, a veritable hypostasis which will be prayed to and invoked, and which will communicate itself to the mind in an ineffable communion.'

The *Yogacara* system attributed to Asvaghosha, but really developed by men like Vasubandhu, Asanga, Dignaga, and Silabhadra in the fifth and sixth centuries, evolved the notion of a subconscious substratum of phenomena, called *alaya-vijnana*, literally sensation of the groundwork. The concept had very great influence on Chinese Buddhism thanks to Hsuan-tsang, who studied under Silabhadra, and on his return from his fruitful pilgrimage, wrote a book on Absolute Idealism in which he expounded the *Yogacara* metaphysics at considerable length.

There are also less ingenious answers. But Gautama himself, who was probably wiser than these Mahayanist metaphysicians and mystics, observed what his biographies describe as 'a noble silence' on the matter. He made no quixotic attempts to pierce the 'sleep' which hangs round our 'little life': to see through some of the dreams which men dream within the dream, was as much as he could hope to do in a life-time.

PART IV

AND THE WORD WAS MADE FLESH

or

SOME ASPECTS OF BUDDHIST ART

THE mystics of Ephesus who edited the Johannine Gospel were men of profound insight: they knew the working of the Word. In the heart of the Word there is an irrepressible urge to become flesh. The Buddhist art of India reveals something of the nature of this intricate and universal process: it shows the Word becoming flesh almost with vengeance.

If renunciation in one form or another was the keynote of the original doctrine of Buddhism, then in the Buddhist art of India we witness the world-renouncers returning to the world by devious paths; returning, if not actually in body, at least in imagination. In body, too, sometimes! There is at Ajanta a fresco—not very important from an aesthetic point of view it is true, but nevertheless possessing considerable psychological significance—which depicts a Bhikhu standing expectantly at the palace door. Mr Laurence Binyon sees in him a spiritual presence, a welcome messenger bringing with him a soft breath of other-worldly peace and serenity into the feverish world of men. He may be right; it is no doubt all a matter of interpretation. But in India we have been accustomed to the world-renouncers sufficiently long to know that their visits to this unsatisfactory world of ours are usually actuated by a purpose other than the mere desire to minister to our spiritual needs. The Bhikhu standing at the palace door may hold the Lotus of the Good Law in his hand. It is evident, however, that he has come not only to preach, but also to beg. In spite of his half-averted, downcast eyes there is, as it were, a certain air of expectancy about the unworldly visitor in the picture; obviously, he expects to be taken in, to be given a meal, since even 'a spiritual presence' has need of 'solid food.' And there is perhaps a still subtler motive to account for the Bhikhu's presence at the gilded threshold of the palace; there is in those eyes a shy desire, a half-conscious, inarticulate desire; and one cannot help feeling that the world-renouncer has come to the palace to have a glimpse of the world he has renounced. For though they could not confess it openly, it seems that these Buddhist monks loved the world; loved its joys and delights, even its pains and griefs.

The art of Ajanta is authentic and genuine; it lives with a life of its own, even though time has dimmed some of its brilliance. From a purely aesthetic angle, the pictorial tradition of Indian art here seems to touch its zenith; already at Ellora, as Dr Coomaraswamy has observed, the representation begins to lose its living quality and tends to flatness; and ever since, there has been a steady and distressing decline till, in our own days, Indian painting is hardly distinguishable from the inanities of the picture-postcard industry. But at Ajanta there is vision at work, with the result that even the faintest figure that can still be seen emerging from the colourless clay, has a grace and beauty which belongs to the highest art. Using the simplest colours—lamp-black, red ochre, yellow ochre, and lapis-lazuli—the painters of Ajanta succeed in realizing delicate and vibrant tones which endow their forms with a movement and animation that is convincingly of the earth and flesh. Their approach to composition is spontaneous, but a spontaneity which points to full awareness of the problem involved, and not carelessness. Their grouping might appear crowded and complex at times, but there is seldom incoherence or chaos. The arrangement of figures, the balance of movement and attitude, partake of a large rhythm such as is inherent in life; and which, though it is not insisted upon by means of theatrical gestures as in the canvases of some of the Renaissance and post-Renaissance painters, is none the less real and palpable. As M. Ivan Stchoukine remarks with keen perception, 'The abundance of forms is far from producing

LISTENING TO MUSIC
(attributed to Chou Fang. From 'Introduction to Chinese Paintings' by Arthur Waley—*Ernest Benn*)

chaotic designs, opposed to all ideas of unity; used with art, it helps to create well-planned compositions, the unity of which is not impaired by the fact that it is made up of heterogeneous elements.' The perspective, as in the early Italian frescoes, is empirical, but it is never arbitrary, and there are no descents into the grotesque.

On the whole the treatment of subjects is far more refined and urbane than anything we meet in late mediaeval Christian art. There is, in fact, a distinct suggestion of depth, of relief even, which is in some ways quite as satisfying as the western tradition of representation that tends to resolve pictorial problems in terms of the plastic art, and which has led us to expect statues in paint. At Ajanta, unlike the Sistine Chapel, one is never in doubt that it is the brush, and not the chisel, that has been at work; and if the Bodhisattva Padmapani lacks the three-dimensional monumental grandeur of Michael Angelo's Promethean Isaiah, for instance, it has the advantage of being of flesh and blood, and not stone.

These, however, are technical and aesthetic subtleties. The significance of Ajanta derives from the fact that, like all authentic and genuine art, it represents a kind of apocalypse: it reveals in its full urgency a conflict which seems to be fundamental to Buddhism in general, and Buddhist monachism in particular. True, the avowed aim of the monks who painted these frescoes was to illustrate the legends of the Great Life for the edification of the faithful. True, they had no conscious intention of expressing the dichotomy of their own souls, and they would have been extremely shocked if any one had imputed such a profane motive to them. But all art, whether religious or secular, involves, and cannot help involving, some sort of a personal confession on the part of the artist. The artists of Ajanta, for all their attempts at impersonality, actually tell us more about their own state of mind, their own struggles and vacillations, than the experience of the Buddha. That is what gives Ajanta its importance.

Ajanta impresses. Firstly, there is the scene; desolation alternating with luxuriant verdure; and Waghora tumbling into the ravine in a series of seven cascades. The caves are cut into a precipitous rock and arranged in the form of a crescent. Inside the caves there is twilight—a twilight that invites speculation. What, one wonders, were the thoughts which possessed the Buddhist cenobites who dwelt in these caves, unconcerned with the Mundane Dispensation? Did they spend their time contemplating the various stages of the Eightfold Path and the corresponding Beatitudes? Did they meditate upon the meaning of the Four Noble Truths, upon the ineffable joy of arahatship and Nirvana? Were their dreams centred round the Foundation of the Kingdom of Righteousness? Were they always reflecting on the doctrine of *Sunyata*, which, 'in the theory of a world as will and representation . . . is the state of a mind free of both representation and will?' Were they pre-occupied with the pursuit of the elusive *alaya-vijnana*, or the Universal Subconscious?

From the walls and ceilings of these caves we receive the answer to our questions; but it turns out, alas, to be a negative answer in every instance. The forms which loom out of the twilight of Ajanta leave little room for vague speculations. The spirit of renunciation speaks for itself; and, apparently, it speaks of very different things from those which one is in the habit of associating with it. Above everything, the world-renouncers of Ajanta appear to be obsessed with the mystery of female flesh and its promises of bliss; they crave for its tenderness and comfort with an almost adolescent passion. Fluttering against their mental eyes are visions of dancing girls, of tender-limbed maidens reclining in inviting postures of abandon and nakedness, of voluptuous princesses disporting themselves with their youthful lovers. They dream of brilliant scenes at Court, of pot-bellied householders relishing wine with ribald pleasantries, of the thrilling game of dice. They imagine the ecstasy of the lovers' intimate embraces, and an almost audible sigh of regret escapes their lips; all the Beatitudes of the Path, even Nirvana, seem cold comfort beside the solid certitude of conjugal felicity.

It is easy to choose; difficult to abide by one's choice. And there are always regrets. Reassuringly, the Bodhisattva Padmapani twists the Lotus of the Good Law between his beautiful fingers; reassuringly, but he does not succeed in giving comfort. In the faint background there is the tantalizing suggestion of flesh; a woman seems abandoning herself to her lover; there are heavenly musicians playing on harps; and finally, as a symbol of the whole mood, there is a peacock crying to its mate.

The Good Law will perhaps never become an adequate substitute for the consolations of flesh, no matter how long those delicate fingers twist the stalk of the chaste lotus they hold. It is difficult to renounce the world; difficult to break away from a thousand subtle psychological ties which bind one to it. One may put on saffron robes; one may take upon oneself the strictest discipline; one may retire into solitude and seclusion. But there is still the imagination. It cannot be put into a monastic uniform; no cloistral walls are thick enough to hold it prisoner; it is impervious to all vows of chastity; it can never be secluded, because it peoples its solitude with creatures of its own making. It is bound only to one thing—desire. And so the world-renouncers invariably return to the world and to the flesh—if not in body, at least in imagination. Such is the unequivocal answer of Ajanta. And it finds an echo across the Himalayas in the grottos of Tun-huang, and at Lung-mên, and beyond the coastline of the Middle Kingdom, at Nara, and way down in the Southern Seas—at Borobudur.

Mr Laurence Binyon, whose views are always illuminating, has observed: 'To Italian painters of Quattrocento, to Pollaiuolo and Signorelli, and following them the young Michael Angelo, the naked body was a discovered romance.' This seems to be true also of the sculptors of Sanchi and Amravati: for them, too, the naked body was an ecstatiating discovery and revelation. There are, of course, important differences of technique and medium. But when due allowance has been made for these obvious, and ultimately insignificant differences, one has to admit that one experiences in their statuary and reliefs something of the same 'thrill of romantic strangeness in the naked forms of joy' which Mr Binyon keenly senses in the nudes of Signorelli.

It is true enough that nudity has never been an uncommon sight in India, though even here Mr Binyon tends somewhat to exaggerate the advantage which the Indian artist is supposed to have had over his western confrere. However, in so far as Buddhist art is concerned, this argument has but little relevance. The Buddhist monks were not normally expected to while away their leisure in loving contemplation of 'the beautiful ways of the human body in its spontaneous gestures and poses.' In fact, ever since Visakha, the rich and prudish widow of Savatthi, had been shocked by the spectacle of Bhikhunis bathing naked on the river side by side with the unchaste courtezans, and Bhikhus tormented by heat exposing their bodies to rain, both male and female mendicants had been enjoined to observe the utmost circumspection even with regard to their own bodies. Primitive Buddhism, for reasons not difficult to understand, had developed into a strict puritanism; it had set up an impossible ideal of celibacy; and, especially in its monastic form, it had inculcated the same horror of the human body which we find in the Christian monachism. Not only was there no question of the Buddhist monks being allowed to feast their eyes upon 'naked glory,' but they were actually expected to look upon the body as 'a sore with nine openings.' In particular, they were told to avoid all contact with female flesh; the Buddhist elders had gone to the extent of introducing into Gautama's deathbed utterances what seems on the face of it an absurdly irrelevant and spurious conversation: 'How, Lord, are we to conduct ourselves with regard to womankind? Don't see them, Ananda. But if we see them, what are we to do? Abstain from speech. But if they should speak to us what are we to do? Keep wide awake.'

This recoil from, and disgust with, things carnal is not confined to primitive Buddhist monachism. It is present in practically all its subsequent developments, and, indeed, ends by becoming an integral part of the Buddhist psychology. Thus in the seventh century we get Santideva, whose *Journey towards the Light* enjoyed among fervent Buddhists something of the popularity of Thomas à Kempis's *De Imitatione Christi*, exclaiming (in M. Finot's translation): 'Consider this corpse, dragged hither and thither by greedy vultures. Why does it offer no resistance? Why, O my heart, dost thou watch over this mass, taking it for thy ego? . . . No doubt this vile body is for men an instrument of action. But thou guardest it in vain, pitiless death will seize it from thee to throw to the vultures. Then what wilt thou do? . . . When shall I go to the charnel-house, the fit dwelling of the body, to come into the presence of the corpses of others and of my own body destined for corruption? . . . Behold my body, behold the decay it will become; its odour will drive away even the jackals.'

Yet recoil is merely a form of attraction; disgust a kind of fascination. The Buddhist recoil from, and disgust with, the body has always been counterbalanced by an equally powerful fascination and attraction towards it. Even Santideva could sing in eloquent strophes of the exquisite figures of 'thousands of *apasras*' who welcome the Manjusri to heaven; strophes, which, as M. Grousset remarks, 'might serve as a commentary on the frescoes of Ajanta.'

This element of attraction and fascination is manifest in the statuary and reliefs of Sanchi, Bharut, and Amravati, where we witness the unknown Buddhist masters positively revelling in the naked forms of joy. It was probably the first opportunity they had had of experimenting with the possibilities of the human figure and its nakedness; for prior to the building of these stupas, the rule laid down in the *Cullavagga*—which restricted the decorative scheme of monasteries and shrines to designs of 'wreaths and creepers, and bone hooks and cupboards' and forbade 'imaginative drawings painted in figures of men and women'—was apparently observed with some strictness. Thus to the sculptors of Sanchi, Bharut, and Amravati the 'naked glory' was as much a symbol of emancipation as it was to Pollaiuolo, Signorelli, or the young Michael Angelo. One has only to look at the moulded contours of the *Lay-Worshippers* from Amravati to perceive a measure of the ecstasy which the world-renouncers experienced in the creation of those voluptuous curves. Somewhere in *Jesting Pilate*, Mr Aldous Huxley writes that Indians will never make good artists, because they are too much interested in metaphysics and ultimate Reality. His generalization is so completely at variance with fact, that it hardly requires contradiction. It might be pointed out, however, that anything less metaphysical, less concerned with the ultimate Reality, than the back view of the *Yakhshini* from the East Gate of Sanchi, it is difficult to imagine. Greek sculptors are by no means unique in having left behind them lasting monuments to the charms of Kallipygos: there are too the unknown Buddhist masters.

For Baumgarten, aesthetics was 'the science of sensuous knowledge.' For the artists of the classical age of India it was something even more precise. No matter what the Silpa Sashtaras, or the accredited interpreters of Indian art might say, the Indian artists of the classical period regarded aesthetics as the science of sensual knowledge. They might be trying to represent 'the ultimate reality,' but they always had the sense to represent it in a form natural to their medium. Even Mr Aldous Huxley has had time to correct his early generalization on this point. Contrasting the almost algebraical airiness of Maya sculpture to 'the suffocating animal heat' of Hindu art, he says in *Beyond the Mexique Bay*: 'From Cape Comorin to the Himalayas, and for the last two thousand years, almost every Hindu artist seems to have been engaged in illustrating the works of Aretino. Even the most sacred persons tend to melt—and at the most solemn moments of their religious life—into suggestive postures . . . The boneless limbs—dozens of them, very often to a single personage—ooze about the picture-space like voluptuous ectoplasm. The haunches jut to right or left; the waists are tapered as though by a delicious process of suction; even the men seem as though inflated about the chest; and as for women . . . But language fails.'

The sensual preoccupation which, with mingled feelings of disgust and enchantment, Mr Huxley notes in Brahmanical art, is also present in Buddhist art. Present, but with a difference. Whereas Brahmanical art is sensual in a straightforward and direct manner, the sensuality of Buddhist art appears to be invariably regressive. The difference becomes obvious if we compare the Vaishnavite and

Saivite sculptures of Khajuraho and Konarak with those of Sanchi and Amravati. They point to two very different forms of sensuality. In the one instance we have sensuality—in the case of the Saivite art, one might almost say sexuality—stretched to the limit of erotic mysticism; in the other, the sensual obsession has its origin and culmination in a nostalgic desire—a nostalgic desire, which, because it recognizes the impossibility of any consummation, trembles on the verge of pathos and languorousness.

The nostalgic quality of Buddhist art emerges even more clearly in the catacombs of Ajanta. For those amorous couples whom we see dallying in shady alcoves behind the Bodhisattva do not strike the almost monstrous note of triumphant phallicism which is of the essence of Saivite art. They are not sensual even in the sense in which the nudes of Titian and Giorgione, or Rubens and Renoir are sensual. Their sensuality is more that of Gauguin in such pictures as *Nave Nave Moe* and *Noa Noa*. There is about them a strangely elusive quality which can best be described as symbolic. They are not just forms and images taken from the tangible world and presented in terms of pure equivalence: they seem to be symbols—symbols of the nostalgia of the disembodied spirit for the warm assurance of the world and the flesh. And in a sense they are not tangible forms at all, but merely phantoms and spirits, haunting the lovesick imagination.

The Mahayana was literally a Great Vehicle. In the course of what is known as 'the Flight of the Mahayana' very unusual things happened. It was, of course, an unusual flight. It led Nagarjuna into the bosom of absolute vacuity. It took Vasubandhu to the edge of the Universal Subconscious. It set up the ideal of an abstract Buddha State, which, according to Asanga, 'is like a shrine of precious stones, great in power.' It made Harsha, the poet king, go into orgies of charity. It sent Bodhidharma from the Coast of Coromandel to Canton. It induced Fa Hsian, the Master of The Law Hsuan-tsang, I-Ching and many other Celestials to undertake hazardous voyages to the Buddhist Holy Land. It created a superb 'Mystic Heaven' of whose lures Santideva has sung in such eloquent terms in his *Journey to the Light*. In this flight, M. René Grousset tells us, 'the external world had dispersed like a dream.' But, perhaps, it is not so easy to disperse the external world from the scheme of things. Li Shang-yin may call the world 'small as a grain of dust'; the Master of the Law may speak airily of the Absolute Idealism; and their best modern interpreter, M. Grousset, may dilate on the infinite potentialities of the *tathata*. But it is not possible to explain away the world in such a summary fashion. Mirage or reality, it cannot be dispersed merely by verbal incantations. Even M. Grousset is aware of the difficulty. After taking us into an 'ocean of clouds,' and solemnly telling us that 'all the rest is only appearance and dream,' he declares rather quizzically 'but the finest dream of all is, that all remains.' All remains because actually nothing had ever been dispersed by the Mahayana metaphysics. It was all a kind of self-hypnosis. For if the testimony of Buddhist art is to be believed—and it is an authentic enough testimony—then the truth seems to be that the flight of the Mahayana had ended before it had begun. It had ended, shall we say, in the Kallipygos of the Dryad of Sanchi.

Who is this mysterious apparition? We do not know her identity, because she appears in different guises in different places. Mr Havell, whose intuitions are seldom wrong, traces her back to the Garden of Eden. Since his fall, man has had opportunity of knowing her better on earth. For she herself is the earth in all its sensuous warmth and fertility. She is a woman, and she is a tree—numinous and yet palpably of the flesh. Here we see her as the spirit of the forest, the wood nymph of whom the Vedic Hymns speak as Arnayani. And the forest is the primeval forest of desire. Beyond Good and Evil, below Good and Evil, she leans forward from the bough of a mango tree, offering herself with a voluptuous gesture. And her breasts are more deliciously succulent than the mangoes hanging from the tree. They play havoc with the spirit of renunciation. In vain does the Bodhisattva Padmapani in those dimly-lit catacombs hold out the blue lotus to the faithful.

Perhaps not altogether in vain. There is also another element in Buddhist art. It cannot be seen in the Buddhist art of India. It is not to be found among the 'Thousand Buddhas' of Tun-huang, with some of whom M. Pelliot and Sir Aurel Stein have made us familiar; nor across the seas in the frescoes which adorn the golden temple of Hôryûji at Nara; and still less in Krom's albums of reliefs of Borobudur. Further, this element does not manifest itself very conspicuously even among the works of avowedly Buddhist masters like Ku K'ai-chih and Wu Tao-tzu; these worthy Celestials appear to be more interested in painting the Western Paradise and the Buddhist Purgatory than in understanding the profounder implications of the Buddha's word. Where, then, must we seek for it?

The answer, of course, is—in China. For the whole of Chinese art and poetry of the T'ang and the Sung periods appears to be permeated with it. Its expression in poetry may be judged from the following poem by T'ao Han:

The pine and the cypress hide the mountain gorge,
But in the West I discover a narrow path.

The sky opens out, a peak is revealed,
And as though it were born in space, a convent rises up before my eyes.
The building seems to be standing on a terrace of cloud.
Its pavilions soar into the air amidst the rugged rocks.
Night comes; monkeys and birds are silent.
The sound of the bells and the song of the bonzes penetrate beyond the clouds.
I contemplate the blue peaks and the moon which is mirrored in the waters of the lake.
I listen to the sound of the streams and the wind that tosses the leaves on the banks of the rushing torrent.
My soul has flown up beyond things visible
Wandering and captive at once . . .

This is not merely verbal mesmerism: it signifies a new quality of vision. And this new quality of vision makes itself felt with equal urgency in the landscapes of Hsia Kuei and Ma Lin. Indeed, it is present to a greater or lesser degree in all the best sketches of the T'ang and the Sung epochs.

What does this new quality of vision imply? Definition in such matters is difficult, generalization hazardous. But it seems that here we see the world and its phenomena—whether they happen to be precipitous mountains, or solitary hermits in their dark retreats, or boats floating on moonlit waters, or birds sitting on fragile boughs, or reeds agitated by the wind—through a consciousness which is no longer time-haunted and time-weary, because it has understood the very nature of time. Here we are in a universe which is devoid of tension—not because contrarieties and conflicts have ceased to operate, but because they have somehow become intelligible. Here, in the very contemplation of transiency, we receive a measure of eternity. Here the mind manages to realize a stasis within the flux itself, so that even a waterfall—normally symbolizing incessant movement—emerges as a symbol of stillness. Here the human soul is at once 'wandering and captive.' Here the wheel turns and does not turn. Here the paradox is no longer a paradox, but rather a luminous certitude. Here we are in the very heart of peace.

PART V

THE BUDDHA IN A CHANGING WORLD

Il y avait une fois La Réalité.
Louis Aragon

IN this vortex of change and becoming which is our world, even the person of the Buddha—the pivot of the Triple Jewel of Buddhism—could not be expected to remain unchanged. 'The Diamond Throne of the original enlightenment,' writes Okakura Kakuzo with characteristic oriental floridness, charm, and profundity, 'is now hard indeed to discover, surrounded as it is by the labyrinths of gigantic pillars and elaborate porticoes which successive architects have erected, as each has added his portion to the edifice of faith.' From the earliest times to the present day mythographers of varying degrees of talent and ingenuity have subjected the Tathagata to their well-meant, if somewhat embarrassing, attention. His apotheosis, in spite of his repeated warnings, commenced during his own life-time. As we have seen, all his logic and reasoning was unable to restrain the enthusiasm of the faithful. They lost no time completing the process of apotheosis at the first available opportunity—that is, almost immediately after the Blessed One had been cremated and his ashes distributed among a number of cantankerous nobles, anxious to make sure of their own salvation by building *dagabas* over his remains. Although Gautama himself professed no knowledge of anything beyond this world, and refused to dogmatize over the first and last things, those who have followed in his wake have taken great pains to work out a richly intricate eschatology for his benefit, which, for the sheer complexity of its invention, would do credit to the imagination of a Dante, or the producer of a modern Grand Guignol. Out of a spirit of revenge, as it were, he has been installed as the impresario of something like seventeen exquisitely delicious paradisiac spheres on the one hand, and saddled with eight blood-curdling 'principal' hells (to leave the innumerable minor hells out of our reckoning) on the other; and what is more, the authors of this colossal scheme of heavens and hells have crowned their remarkable achievement, so Lord Zetland observes, by finding room for it 'within the categories of space and time.' A self-conscious and unassuming intellectual, the apocalyptic fantasy has transfigured Gautama into 'the Great King of Glory'—a rôle which he was never fitted to fulfil, and which he would have found extremely distasteful had he been compelled to play it. A tragic philosopher who could be accused of almost anything but harbouring immortal longings, the hagiographers have had no scruples about crediting him with comforting notions of personal survival.

All these and many more ironical developments of historical Buddhism were no doubt inevitable. Two thousand years is a long stretch of time; and it is natural that during this period a formidable and fantastic superstructure of myth and legend should grow round the figure of the Buddha. The process of this growth makes a most fascinating study in religious architectonics, which, however, is beyond the scope of the present book. It is sufficient to observe that the process of Buddhist diffusion, a process which has, to some extent, influenced practically every culture flourishing in the Asiatic world, has been far from an unmixed blessing. Thomas Huxley in his Romanes lecture found some hope in the fact that 'a system which knows no God in the Western sense, which denies a soul to man, which counts the belief in immortality a blunder and the hope of it a sin, which refuses any efficacy to prayer and sacrifice, which bids men to look to nothing but their own efforts for salvation, which in its original purity knew nothing of vows of obedience and never sought the aid of the secular arm: yet spread over a considerable moiety of the old world with marvellous rapidity and is still with whatever base admixture of foreign superstitions, the dominant creed of a large fraction of mankind.' But, in honesty, it has to be admitted that it is not by virtue of any of its more pleasant features that Buddhism managed to 'spread over a considerable moiety of the old world with marvellous rapidity'; rather, it was because of its extreme readiness to welcome into its folds every foreign superstition with which it came in contact. To-day 'the Diamond Throne of original enlightenment' accommodates not only the lean Tathagata, but also the vast army of the demons of 'a considerable moiety of the old world'; and the 'edifice of faith' is haunted by ghosts whose presence is a menace to human sanity.

Thus the mere statistical fact that 'the present total of Buddhists in the world is about five hundred millions, or twice as many as that of the Christians—Roman Catholics and other sects included,' hardly affords any solace. Numbers mean little in themselves. In the present case, indeed, the statistics only serve to bring out a tragic fact. H. J. Massingham in his penetrating essay on Buddhism comes to the conclusion that Buddhism in its development as a world-religion has been a 'tragedy.' And he is right. The growth of Buddhism as a world-religion shows how utterly impossible it is for human beings to accept truth in its objective simplicity; it shows, too, that they cannot be content with the principle of reason alone for any length of time; and finally, it shows that to integrate themselves at all in relation to the bewildering but beautiful universe around them, they must resort to the tragic mediacy of the Myth. Among many becoming, honorific titles which were bestowed upon Gautama by fervent piety, one which has always enjoyed a very wide currency is 'the Light of the World.' The title has only a pathetic interest: 'the Light of the

World' seems to have proved singularly inadequate in dispelling the encircling gloom of human ignorance and stupidity.

Gautama is not merely a dim figure silhouetted against the twilight of India's legendary past. He has a contemporary interest. To-day his influence extends far beyond the boundaries of the mystic East. Curiously enough, the insidious smile of the World-Renouncer of Kapilavastu has cast even over the supposedly materialistic Western world an enchantment akin to that of the Saviour of St Sulpicery fame, who has served as a solace for so many broken hearts. Indeed, as a decorative object pure and simple, the image of the Enlightened One possesses an advantage over that of the tortured Galilean nailed to the Cross: the former is more soothing, more restful. It is, of course, a far cry from the primitive mints of Maues, where according to Dr Coomaraswamy's researches, the Buddha image, as we know it, made its début some time in the first century B.C., to the highly efficient emblem-manufacturing factories of Birmingham which thrive on the love of the faithful. It was to be expected that in the course of this long journey through space and time, the original image should have undergone radical transformations. It is, however, a matter for gratification that these changes have not meant any loss of its magical appeal. In fact, if anything, they have increased it.

They have increased its appeal by the simple fact of making the image accessible to a far larger mass of people than had been possible at any time before in the history of Buddhism. Evidently, it needed a democratic age to turn out luminous statuettes of the 'Self-Luminous' One in sufficiently large numbers at prices within the reach of his humbler devotees. The Buddhist missionary-artists of the great age of Mahayana could only cherish, as a pious dream, the hope of providing a Buddha for every Buddhist home; they struggled in vain with the hopeless limitations of their technique, which rendered the realization of their ideal impossible; there are, alas! mountains which even the most ardent faith cannot remove. But the modern image-makers —those invisible feeders of faith—with their wizardry of mass-production have succeeded in working the miracle; the Face of Silence can at last be seen in practically every curiosity shop east as well as west of Greenwich; serene, detached, slightly cynical and world-weary, it seems to watch the mad rush of this age with a baffling indifference.

Realized ideals, however, often bring disillusionment in their trail. In the present case, too, there has been cause for disappointment—indeed, even indignant complaint. Mr Will Hayes, a writer on diverse religious topics, has recorded some instances of minor tragedies arising from the extensive diffusion of the Buddha image in an industrial age. Not long ago, he tells us, the Buddhists in England had to seek the intervention of the authorities with regard to a 'Buddha Hat-Pin,' and also were compelled 'to ask one of the big bazaars not to sell a sixpenny trinket made in the shape of their Master.' He further speaks of a personal experience. 'The other day,' he writes feelingly, 'I discovered in a shop window a figure of the Buddha labelled "Buddha Paper Weight." ' Had Gautama himself seen this, perhaps he would have enjoyed the joke. But one cannot be a *Bhakta*, a devotee, without dispensing with one's sense of humour: the author of 'My Buddha,' whose deep attachment to his hero is unquestionable, was naturally chagrined rather than amused. 'I walked straight into the shop,' he continues, 'and gave the young lady in charge a serious little lecture.' The young lady in question was unable to see her transgression until she was asked 'what she would think if she looked in the window across the road and saw

"Jesus Pepperpot" for sale.' Quite justifiably indignant, Mr Hayes wags an admonitory finger at the 'Christians who have to realize that these things are not in good taste.'

That these things reflect a deplorable lapse of taste on the part of Christians, is true enough. But although one may sympathize with the British Buddhists in their great tribulation, it has to be recognized that the application of Noble Truths *en masse* was bound to create awkward situations like these. It is not known whether the protests led to the withdrawal of the offending articles from the market. But on the face of things, it is unlikely that mere consideration for the hypersensitive susceptibilities of a few votaries of Buddha would induce the souvenir manufacturers to abandon a very lucrative item on their list. Perhaps before long we shall get used to the idea of Buddha hat-pins, of cheap trinkets and decorative paper-weights made in the shape of the Great Master. After all, the image of Buddha has been subjected to far worse indignities in some of the Buddhist lands themselves. To take an instance which is familiar history, we might mention the story of the infamous king Bidatsu of Japan. The fact that he had been a devout and practising Buddhist did not prevent him from having the temple at Yamato burned, ordering the Buddhist nuns and monks to be flogged, and having the sixteen feet high image of the Buddha thrown into the canal; all this because the Buddha had failed to deliver his people from a severe epidemic of smallpox which had broken out in his kingdom. Such misfortunes must be regarded as an essential part of the dialectics of apotheosis. Those among men who are raised to the pedestal of godhead must also suffer the fate of gods, who are always in imminent danger of being treated as scapegoats.

The word of the Buddha first reached the West by the

overland route. One of the edicts of Emperor Asoka, the great patron of Hinayana Buddhism, who has been sometimes compared with Constantine, makes a claim on his behalf that he propagated the Dhamma as far as the kingdoms of the four kings named Ptolemy, Antiochus, Magas, and Alexander, who are now believed to have ruled over Syria, Egypt, Macedonia, Cyrene, and Epirus. The claim is probably based on fact; for after the Alexandrian invasion, contact between India and the West was almost continuous, and trading colonies of Indians are known to have existed at various important centres on the coast of the Mediterranean, the one at Alexandria being well-known. Some time in 21 A.D. Augustus, while on a visit to Athens, received an embassy from India which had come by way of Antioch. According to Strabo, there was attached to it a man named Zamanochegas. Sir Hari Singh Gour suggests that this is possibly a corruption of Saman Achary, and that he was probably a Buddhist bhikhu. The hypothesis seems credible, and offers an explanation for the extraordinary behaviour of the said Zamanochegas: he provided the Athenians with an unusual sensation, burning himself alive in public. It is not known whether the Athenians were impressed by his self-martyrdom; but there is no doubt that this was probably the first demonstration of the doctrine of renunciation carried to its logical extreme.

Towards the closing years of the second century the name of the Buddha actually appears in the writings of a Church Father, wrongly spelt it is true—but still unmistakably his name: Clement of Alexandria writes of 'those of the Indians that obey the precepts of Boutta, whom through exaggeration of his dignity, they honour as god.' Al Biruni's quotations from *Shaburkan*—a work by Mani, the great heretic of Ecbatana (215-276 A.D.)—leave little room for doubt as to the fact that Manichaeanism was profoundly influenced by the Buddhist doctrines; indeed, Mani claimed to be in the direct line of Buddha, Zoroaster, and Jesus. References to Buddha in the *Acts of Archelaus* and St Jerome have already been mentioned. In the thirteenth century we have Marco Polo representing the Buddha as the son of a king of Ceylon, and sadly reflecting what a pity it was that Sagomani Barcan (Marco Polo calls Gautama by his Mongolian title) was an idol-worshipper and not a Christian, or else he would have made such 'a great saint of our Lord Jesus Christ, so good and pure was the life he led.'

The Anglo-Saxon world first came to know of Buddha through an adventurous sailor Robert Knox (1660) who has something to say about 'a great god, whom they [meaning the Singhalese who had kept him a prisoner for nineteen years] call Bouddou, to whom the Salvation of Souls belongs'; and he goes on to add in *An Historical Relation of Ceylon*, the fruit of his captivity, 'Him they believe once to have come upon the earth. And when he was here, that he did usually sit under a large shady Tree, called Bogahah.'

It was not until the nineteenth century, however, that any attempt was made to obtain real knowledge of the Buddhist sources, though Simon de La Loubère, the representative of Louis XIV at the court of Siam (1687-8), had tried to translate a few passages from Pali books. The first great discovery of the Buddhist scholarship was made by Alexander Csoma de Körös, a Hungarian scholar; he was trying to establish the origin of his people when he accidentally came upon some Tibetan Buddhist texts at Calcutta. Meanwhile, Brian Houghton Hodgson, for ten years (1833-43) the British Resident in Nepal, had gathered over 400 Sanskrit MSS. The discoveries of Csoma and Hodgson were to form the groundwork of Eugène Burnouf's *Introduction à l'histoire du Buddhisme Indien*, which appeared in 1844. It was a monumental work, but its appearance was somewhat eclipsed by the fact that almost simultaneously George Turnour had unearthed the Pali sources; for, as was later to be proved, the bulk of Burnouf's sources were much later than the Pali.

During the latter half of the nineteenth century a most impressive structure of Buddhist scholarship grew up through the works of Victor Fausböll, Herman Oldenberg, Thomas William Rhys Davids who founded the Pali Text Society, Spence Hardy, and Senart. This tradition of painstaking scholarship has been maintained in our own times by men like T. Stcherbatsky, Louis de La Vallée Poussin, Suzuki, E. J. Thomas, and E. L. Woodward, to mention only a few.

But the Buddha who has captured the popular imagination of the West is not at all the Buddha who is to be found in the works of Buddhist scholars. Popular imagination likes unanimity; and scholars seldom find it possible wholeheartedly to agree on any subject. Popular imagination wants to believe; and acquaintance with works of Buddhist scholarship still sometimes tends to lead to the scepticism which the Russian scholar Vasiliev expressed, almost a century ago, saying: 'Russian, French, English, and German scholars have in fact written much on the subject. I have read through most of their works, but through them I have not learnt to know Buddhism.' So the popular mind in the West, as in the East, has accepted the Buddha of whom Robert Knox speaks: 'the great god to whom the Salvation of Souls belongs. Him they believe once to have come upon the earth. And when he was here, that he did usually sit under a large shady Tree . . .' The attraction of the Bodhi-Tree is apparently irresistible; its shadow extends even across scholarly pages.

'During the late nineteenth century,' J. B. S. Haldane

writes in his *Fact and Faith*, 'certain oriental religions became temporarily fashionable in "advanced circles."' That fashion is still with us; and we must attribute to it the curiosity which the Buddha and his doctrine have aroused at varying levels of response, from the highest downwards. The popularity of some of the oriental religions among 'advanced circles' is easy to understand; the popularity of the Buddha and his doctrine, however, raises some difficulty. For the idea of Renunciation at some point in a man's psychological development seems to be essential to any attitude which can legitimately claim kinship with the Buddhist view of life. This idea is, of course, capable of being interpreted in a variety of ways; and the motives leading to it will always reveal the widest divergences. But even taken at its crudest, that is when it is indistinguishable from various other doctrines of compensatory wish-fulfilments in a fantastic hereafter, and is a kind of *quid pro quo* for establishing 'proper connections with the higher powers,' it still remains a puzzling anachronism in a world frantically preoccupied with the adoration of Mammon.

Quoting Renan, who maintained that one could not be polite in a Paris omnibus without breaking the rules of the company, Professor Irving Babbitt in the introduction to *Being Creative*, observed with a quiet wisdom that characterized practically everything he wrote: 'Even so, any one who sets out to be a Buddhist to-day would find himself in conflict with the underlying assumptions of our civilization.' This is undeniable. And yet, it is also a fact that it was precisely at a moment when these assumptions—which Irving Babbitt recognized as being fundamentally in conflict with the essential Buddhist point of view—were becoming fully crystallized that the popular interest in Buddhism was greatly stimulated. The age which saw the translation of the Sacred Books of the Buddhists, also signalled the emergence of a chauvinistic and shamelessly venal Imperialism; it is now seldom remembered that Max Müller and T. W. Rhys Davids were contemporaries of Meredith Townsend, Rudyard Kipling, and Cecil Rhodes.

Contradictions of this nature have, of course, always existed: there is a ubiquitous incongruity between what human beings admire in the abstract and their actual practice. Julien Benda remarks somewhere that for two thousand years humanity has done evil, but honoured good: his statement perhaps possesses a wider application than he claims for it. Moreover, we always interpret our ideals in such a manner as to make them conform with our desires and interests: religion after all, as William James was to observe, revolves on the interest of the individual in his private personal destiny, and constitutes 'a monumental chapter in the history of human egotism.' This is doubtless quite natural and normal. But it should serve as a warning against that easy optimism which sees in the rise of 'Mahabodhi societies for the study and spread of Dharm . . .' in the Western world an augury of the birth of a 'Third Vehicle,' and a new awakening of compassion in the human breast. Unfortunately, the world is still not fit for Buddhists to live in, and to try to be one is almost as quixotic as trying to-day to be a Christian. It was Nietzsche who said that the last Christian died upon the Cross. A truly Nietzschean exaggeration; but, like most of his exaggerations, it trembles on the verge of a profound truth which no amount of pontifical pomp and eloquence can alter. Similarly we might say that the last Buddhist died of dysentery at Kusinara. Meanwhile, those who like, can console themselves with the pathetic spectacle of saffron-robed Buddhist monks crossing perilous seas and taking their stand in the dazzling glare of modern publicity alongside many other professionals, in the vain hope of establishing the Sangha (Brotherhood) in the West;

of charming Buddhist nuns whose feet have 'even trod Piccadilly.'

The Buddha is a protean character. 'In Earth and Heavens and Hells incomparable,' he appears in different shapes, in different places. He has already appeared on the Indian screen; and Hollywood might yet go one better. M. Axel Raoul Wachtmeister has put him into the operatic repertoire; and though it is unlikely that we shall ever see his Opera-Oratorio at Covent Garden, there is every likelihood that the Buddha may yet appear at the Albert Hall. That the Buddha has pictorial possibilities is proved not only by the art of Ajanta, Borobudur, Lung-mên, and Hadda, but the 'Dictionnaire Infernal' of M. Collin de Plancy: an exhaustive compendium of 'all matters relating to devils, fairies, magic, astrology, etc. . . .'—'There he appears,' says Rhys Davids, 'in a curious woodcut as "Sakimuni, génie ou dieu," in the character of the Man in the Moon; or rather of the Hare in the Moon.'

All these rôles of the Tathagata are easy to understand. But what is surprising is that he appears also in the Roman Calendar. Marco Polo's wish has, in fact, come true: 'Sagomani Barcan' has actually become a Christian saint. Somewhat surreptitiously it is true, but unmistakably he has found his way into the rather exclusive hierarchy of the Holy Apostolic Church. He appears there under the incognito of Josaphat, and has a Church dedicated to him as far west as Palermo. Joseph Jacobs collected the many variations which exist of 'the profitable story brought to the Holy City from the farther part of Ethiopia called India, by John the Monk (probably St John of Damascus), an honourable and virtuous man of the monastery St Sabas, containing the life of Barlaam and Josaphat, famous and blessed men both.' The story enjoyed the reputation of a Church bestseller during the middle ages. It is to be found in Vincent de Beauvais's *Speculum Historiale*, Jacques de Voragine's *Golden Legend*, and was sung by the Troubadour poet, Guy de Cambrai. It was so popular that it was translated into practically every European language—even Icelandic.

In the eighteenth century 'a Reverend Divine' of London published at the price of 2d., *The power of Almighty God, set forth in heathen's conversion shewing the whole life of prince Jehosaphat, the son of King Avenerio of Burma in India in seven parts; how he was converted and made a Christian which was the conversion of his father and the whole land*. But it was Max Müller who first drew attention in his essay on the 'Migration of Fables' to the fact that 'all that is human and personal in the life of St Josaphat is taken from *Lalita Vistara*.' In other words, he argued that St Josaphat was none other than the Buddha of the Sanskrit Canon in his initial rôle of a Bodhisattva. 'Thus the Sage of Kapilavastu,' he observed, 'has received the highest honour that the Christian Church can bestow.' If this argument were admitted, November 27th, which is assigned to Barlaam and Josaphat in the Roman Martyrologium, should really be known as St Gautama's day.

Official representatives of the Church, however, take a different view of the matter. Vicarious atonement may be permissible, but they are not prepared readily to admit the validity of vicarious canonization. Their attitude on this point has evoked a spirited protest from Mr Will Hayes. 'Ignorance of religious history . . .' he complains bitterly, 'is one of the average Christian's great failings. In lectures on comparative religion I have frequently mentioned the fact that the Buddha is a saint both in the Roman Catholic and Greek sections of the Christian Church. And I know quite well that every time I have said this my reputation for veracity has gone down several points.' But saint or not, on the whole it seems that the orthodoxy 'which would fain narrow the limit of the divine government of the world to the history of the Jewish and Christian nations' has treated Gautama with an indulgence altogether absent from its treatment of other 'false and heathenish prophets.' 'One hardly trusts one's eyes,' wrote Max Müller, 'on seeing Catholic and Protestant missionaries vie with each other in their praises of the Buddha . . . Indeed, no better authority can be brought forward in this respect than that of a still living Bishop of the Roman Catholic Church . . .' The Bishop in question, Bishop Bigandet, is no longer living, but his example has not been lost, and the agnostic of Kapilavastu continues to elicit frequent episcopal tributes.

Now it need not be denied that some of these tributes are genuinely inspired and disinterested; and they indicate, as in the case of the late Bishop Gore and the present Bishop of Birmingham, the growth of a healthier outlook among the Christians. But the motives for praise are not always disinterested. Certain dignitaries of the Church seem to admire Buddhism because they find that in some of its forms, it has constituted itself into a notoriously rigid sacerdotal institution which bears a striking resemblance to Christian orthodoxy. The ceremonials of Lhasa are probably the nearest approach to the glory of the Vatican to be found east of Suez; the chief Temple of the holy city has something of the solidity of St Peter, if not its formal beauty; and the person of the Dalai Lama is as near the haloed pontifical figure as one can hope to find more than twenty thousand feet above sea-level. 'Lamaism, indeed, with its shaven priests,' writes Rhys Davids, 'its bells and rosaries, its images, and holy water, and gorgeous dresses; its service with double choirs, and processions, and creeds, and mystic rites, and incense, in which the laity are spectators only; its abbots and monks, and nuns of many grades; its worship of the double Virgin, and of the saints and angels; its fasts, confessions, and purgatory; its images, its idols, and its pictures; its huge monasteries, and its gorgeous cathedrals, its powerful hierarchy, its cardinals, its Pope, bears outwardly at least a strong resemblance to Romanism, in spite of the essential difference of its teachings, and of its mode of thought.'

Tastes differ. Some Christians, on the other hand, have admired the simplicity of the Buddhist faith. Thus Bishop Bigandet, though himself a dignitary of the Roman Catholic Church, admired Burmese Buddhism because 'philosophy, which during its often erratic rambles in search of truth, changes, corrects, improves, destroys, and in numberless ways modifies all that it meets, never flourished in these parts,' and as such had no chance of tainting 'the religious institutions, which have remained to this day nearly the same as they were when first imported.'

Finally there is the all-important missionary interest of Christendom in gleanings from the Buddha's fields. Since the days of the Justinian Code and the Crusades, the Holy Inquisition, and the extirpation of the Albigensian heretics, Torquemada and the gangs of conquistadors, the Faith Militant has learnt better. There are less drastic ways of saving the souls of infidels and heretics than burning, or stretching them on the rack. At the meeting of the International Missionary Council held in Jerusalem in 1928, discussing ways and means of propagating the true faith among the Japanese, Mr K. J. Saunders made some revealing suggestions. He advocated that Christian workers there should co-operate with the free-thinking Buddhists because 'many in this group are as truly Christian already as most Christians; and others are very near the Kingdom.' He added: 'The problem here is how organized Christianity in Japan can definitely win over those non-Church Christians or semi-

Christians and then through them reach the other groups.' After all, there is some purpose behind all this eloquence from the pulpit in praise of the Buddha: it may be an olive branch to 'those non-church Christians or semi-Christians' who are very near the Kingdom, and through whom it may be possible to do something for the tormented souls of the unchristian millions dwelling in Buddhist lands. Whether this ingenious plan is likely to save the tragic multitudes of China and Japan and the Far-East, it is hard to say. One cannot help feeling, however, that the aspects of Buddhism which ecclesiastical circles, for one reason or another, find so admirable, have about as little, or as much, to do with Gautama as the Completed Doctrine with Jesus.

Gautama and Jesus! The Buddha and the Christ! Comparisons of this kind seldom serve any useful purpose; they generally involve too much simplification of human character; and they are at times even odious. But in the present case the comparison has so often been insisted upon that to-day it is almost impossible to avoid. To mention only the more familiar instances, the comparison is implicit in Edwin Arnold's *Light of Asia*, which purports to be 'the Scripture of the Saviour of the World'; Thoreau himself 'named the Buddha beside Christ,' though he was aware that in so doing he was laying himself open to the censure of devout Christians; and a contemporary French writer has called Gautama '*le frère ainé du Jésus Christ*.'

The problem, however, needs some clear definition. In fact, it is not one problem: there are two. Firstly, there is the question whether there are any points of similarity between the Buddhist and the Christian legend. To this question the answer seems to be in the affirmative. The parallels established by Seydel, Van den Bergh van Eysinga, Pischel, A. J. Edmunds, M. Aneski, and others are unmistakable. Some of them—for instance, the story of Simeon, the Temptation,

Peter walking on the sea, and the Miracle of the loaves and fishes, which Garbe regards as obvious examples of borrowing by the writers of the Gospel from the Buddhist legend—are so striking that they cannot be explained purely as cases of accidental coincidence in religious legends. Even if there were no direct contact between India and Palestine in the first century A.D., the possibility of indirect contact through the Persians was very considerable; and there is no reason why Evangelists should not have known the Buddhist legend in some of its Near-Eastern variations. Actually there has been far more diffusion and inter-penetration of legends both secular and religious, current in different parts of the world than is admitted by most people who seem to take a proprietary interest even in legends.

The second question is whether the personalities of Gautama and Jesus, after they have been divested of their obviously legendary and mythological attributes, have anything in common; whether, that is to say, there is any measure of identity between their respective world-views. This is a very complicated question. For it is difficult to obtain any agreement as to which parts of the Buddhist and the Christian stories are historical, and which are not; what pronouncements can be attributed to Gautama and Jesus, and what is to be regarded as the improvisation of later commentators. M. Loisy has caused considerable embarrassment among Biblical critics by suggesting that 'the Sermon on the Mount . . . was never preached (*prononcé*).' A similar, possibly even more devastating case could be made with regard to Gautama's first sermon, 'the Sutta of Turning the Wheel of Doctrine'; indeed, the argument is implicit in the Sun-Myth theory of Senart and Kern. Such being the uncertainty, it is necessary to narrow down the issue. The question that we should ask ourselves is whether the integral personality which emerges from a critical analysis of the

Pali Canon, on the one hand, and from the Synoptic Gospel, on the other, point to any common basis of experience, or resultant attitude.

To this question one can only return a negative answer. Gautama and Jesus were as different from one another as it is possible for any two human beings to be. The difference is not merely, as Canon Streeter has it, 'that the Christ was a carpenter, the Buddha was a prince; they experienced life from different angles. The Buddha was a philosopher; Jesus had the mind of a poet.' This is unduly simplifying the issue. It is by no means certain that Gautama was a prince, though it is true his people were comfortably off. On the other hand, though there is no doubt as to where Jesus' own heart lay, it remains open to question—in spite of Kautsky—whether he personally had any intimate knowledge of poverty; Dean Inge, for instance, rejects the whole notion which makes Christ '*le bon sans-culotte*,' and even goes to the opposite extreme, suggesting that 'Christ and his apostles belonged to the prosperous peasantry of Galilee, a well-educated and comfortable middle class.' And the difference between a poet and a philosopher is, to say the least, rather difficult to define.

A somewhat novel interpretation of the contrast between the personalities of Gautama and Jesus has lately been put forward by those who, presumably, subscribe to what Papini calls 'the exhausting mercantile superstition of our day.' With varying degrees of refinement, they enlarge on Mr Bruce Barton's conception of Christ as 'a precursor of the modern man of business, an apostle of outer action and even as a go-getter.' Against this they assess the Buddha as a typical oriental, a passive dreamer of dreams. The querulous Buddhist, of course, could protest that, as far as action goes, his catechism insists on 'strenuousness' and 'effort' with such vehemence that it might serve as an excellent text for the modern men of action, whose exploits are normally limited to meditations on Stock Exchange reports, and who are wise enough to have their work done for them by paid slaves. Further—since by their deeds shall ye know them!—they might point to the prodigious achievements of their Japanese confrères who, even Mr Bruce Barton would admit, have been anything but dreamy and passive in recent years. To see the difference between two men in this light is to miss the significance of what seems most central in their world-views. The range of our knowledge about them is limited, but one thing may confidently be asserted: that Gautama was as far from being a lotus-eater as Jesus from our glorified slave-drivers of Big Business and High Finance.

Dr Reichelt, a German missionary of very wide sympathies, who founded 'The Society of Religious Friends' in China, informs us that in his conversation with 'really religious Buddhists' he was able to extract from them the admission that 'in the story of Calvary the Bodhisattva doctrine has reached its perfection.' This must have been due to some misapprehension on the part of our 'really religious Buddhists.' For one thing, the Bodhisattva is nowhere represented as bearing his cross through crowded streets. And the reason why he is not so represented is, partly at least, because Gautama was above all a prudent man, like Confucius, and lived the kind of life which could by no stretch of imagination on the part of his followers be interpreted as martyrdom. He scrupulously avoided getting into trouble with the temporal powers, who would doubtless have given him 'short shrift,' as did Pontius Pilate to the 'Messianic agitator' from Galilee. His concession with regard to the admission of runaway slaves, domestic servants, soldiers, etc., may be mentioned as one example of his desire not to get himself involved in any struggle with Law and Order. Because of this prudence on his part, Buddhism has remained lacking in 'the tremendous fact' which forms the nucleus of the Gospel Passion Play—the fact of a supreme martyrdom. But, on the other hand, for this very reason Buddhists have been spared from having their bowels of compassion moved to the point of excruciation by subtle stabs of remorse. On the whole, Gautama seems to have had more consideration for his votaries: they are not reminded of the man who died and suffered for them every time they visit their shrines; nor do their temples everlastingly echo the silent but agonizing refrain of, 'is it nothing to ye, is it nothing to ye, all ye that pass by?' The idea of dying and suffering for others in the Christian sense is, broadly speaking, alien to practically every form of Buddhism. It was certainly alien to Gautama's teaching; he quite frankly told his Brethren that they must not expect that ultimate self-sacrifice of him: for 'by oneself is evil done, by oneself one suffers. By oneself is evil left undone; by oneself is one purified. Purity and impurity belong to oneself; no one can purify another.' There is no vicarious atonement available to a Buddhist; but then, on the other side, he is spared those painful stigmata which were the lot of 'Christ's Poor Little One,' St Francis of Assisi. It would be futile to compute whose is the greater felicity, the Christian's or the Buddhist's!

Environmental differences, it is true, account for the divergences in their public careers up to a point. Gautama was able to avoid a useless martyrdom partly, at any rate, because both his upbringing and the conditions under which he lived were very different from those of Jesus. Given the circumstances with which Jesus was faced, given also his peculiar conditioning, it is arguable that Gautama, too, might (though not necessarily) have ended on the Cross. Fortunately for him, however, India in the sixth century B.C. presented a very different picture to that of the Holy

Land in 27 A.D. This is a plain historical fact which is too often overlooked. It is not necessary here to go into details of the social and psychological characteristics of the world of Gautama; it will be sufficient to emphasize the more obvious features of the milieu in which he found himself. It is certain that it was altogether more peaceful, less overwrought, than the one in which Jesus rose to prominence.

'Christianity,' observes Oldenberg, 'founded its kingdom in times of the keenest suffering, amid the death struggle of a collapsing world. India lived in more settled peace.' Here the horizon was not overcast with grim forebodings of the Great Event which was to destroy the existing order and usher in a new era. It was precisely such anticipations, at once morbid and exalting, which made it possible for the Prince of Peace to appear on the scene with his highly explosive message; indeed, it may be suggested that they forced the issue for Jesus and left him with no choice.

At the time of the Buddha in India, too, there was a great deal of intellectual and spiritual ferment. But it was something quite different from the 'Messianic ferment' brewing in the lands bordering on 'the Great Sea.' The distinction is important to recognize. People in Buddhist India were interested in attaining salvation, release from the turmoils of earthly existence, and heaven, as people always and everywhere have been interested in their destiny beyond the grave, and in making proper arrangements for an uncertain, and hence terrifying, future which is perpetually glaring man in the face. But with them it was a leisurely, and, comparatively speaking, normal pursuit. There was hardly any touch of hysterical violence about it. God was being sought after by mortals in a variety of ways, but the technique of this hopeless quest was almost scientific in so far as they proceeded by the method of trial and error in the laboratory of personal experience rather than mob frenzy in the market place. There was no attempt to force God to come down to earth and found his kingdom.

Indians have always been a patient people; that has been their principal virtue as well as their major vice. The men of the age of Gautama could afford to wait for their salvation. They stood on their heads if they so desired, or calmly contemplated the tips of their noses with a fixed, unmoving gaze; followed the straight and narrow path of the true Brahmanical faith, or quietly deviated into the fascinating by-ways of heresy; practised strictest austerities or indulged in the most sensational ritualistic orgies. Not so in Judaea of the time of 'Our Lord.' If the whole of Asia Minor, the Peloponnesian peninsula, and shores of the Adriatic and the Aegean constituted a spiritual, as well as political, powder-magazine, the fuse was being laid in Jerusalem. Successive sanguinary wars had left behind them a trail of death, disease, and misery. In this delirium of suffering, the problem of 'future,' the problem of God, had assumed overwhelming proportions, had become an obsession. Miracles were in the air; the atmosphere was tense with expectations of a cataclysmic end 'coming from the East and shining to the West.' Prophecy had risen to a culminating pitch of morbidity and exultation; people were literally being stampeded into the penitentiary. 'The time is fulfilled,' the resonant voice of the oracle in flesh and blood came echoing through the darkness of an apocalyptic hallucination, 'and the kingdom of God is at hand: repent ye, and believe the gospel.'

A veritable panic was precipitated in the Holy City when a young visionary from Nazareth arrived on the stage. This strange young man had already beguiled a small group of followers by promising that he would make them fishers of men. The authorities, however, were not having any sedition put across under the guise of a new mysticism, and they had little difficulty in effectively suppressing the Messiah. In Golgotha, prophecy was put to a severe and ultimate test; as could have been foretold, it failed miserably, and the forces of Law and Order emerged triumphant. Tragically the twilight fell on the last, perhaps the greatest of prophets, and the first among the martyrs. The whole episode was a short one, like a meteor rending the darkening sky with its incandescent pain for a brief moment, then as quickly vanishing.

Gautama's course, on the other hand, ran through a long, steady, and uneventful groove: it was unspectacular to the point of being mediocre and pedestrian. He lived to a respectable old age; and though the doctrine he preached was, in its own way, quite as revolutionary as that of Jesus, he was discreet enough to adopt methods of propaganda and argument which were not likely to bring him into conflict with the temporal power or the hierarchical authority. He ridiculed the superior pretensions of priests and patriarchs alike, but he did so with a grace and gentle irony which seldom aroused needless antagonisms. As far as possible, he avoided making angry scenes; and that sense of almost physical exhilaration, which El Greco's Christ seems to derive from driving the money-changers from the Temple, was not known to him. He made no quixotic promises, disappointed no eager hopes; he coveted no kingdoms and suffered no crucifixion.

These contrasts are obvious enough. Underlying them there were, of course, subtler differences of tradition, of the background of civilization in each case. Behind Gautama there were the pantheistic lyricism of the Vedic hymns, the diffuse speculations of Upanishads and the Brahmanical abstractions; contemporaneous with him were the more metaphysical, though as yet probably unsystematized philosophies of several groups of dialecticians. His predilection was towards intellectuality; his training had been that of an intellectual; and though he clearly saw the limitations of intellect, he did not reject it totally, remaining an intellectual to the end. Jesus, on the contrary, had been nurtured in the highly inflammable tradition of Judaism; a tradition which had produced such eminent prophets as Isaiah, Jeremiah, Ezekiel, and Job; a tradition which, as Canon Streeter admits, had no interest in metaphysics whatever. This was to give his passionate nature an even more passional turn. It accounts for what Sir Charles Eliot calls 'the element of the tragic' in his life; it accounts for his 'nervous irritability'; it accounts for his sorrow tinged with occasional fits of self-pity. 'Christ,' observes Sir Charles Eliot, 'is at conflict with what he calls the world. He is angry with it because it will not hear him. He declares that it hates him. The little towns of Galilee are worse in his eyes than the wicked cities of antiquity because they are not impressed by his miracles . . . But the Buddha was not angry with the world. He thought of it as unsatisfactory and transitory rather than wicked, as ignorant rather than rebellious . . . He troubled little about people who would not listen to him.'

Jesus took an altogether more emotional view of life than Gautama. It is impossible to conceive of him as saying: 'All that we are is the result of our thought; it is founded on our thoughts, it is made up of our thoughts'; as impossible as to imagine Gautama supplicating in some Indian Gethsemane, 'Abba Father, all things are possible unto Thee; take away this cup from me: nevertheless not what I will, but what Thou wilt,' or crying, 'My God, My God, why hast Thou forsaken me?' The idea of filial devotion to the Deity, of forsaking and being forsaken, of abject betrayals with a kiss would have appeared rather absurd to the sceptical and matter-of-fact mind of the Tathagata.

The two ways diverge fundamentally. Dr Reichelt's

'really religious Buddhists' could not have read their Scriptures: Christians, after all, are not the only people to betray ignorance of their religious history. The Bodhisattva doctrine is not the doctrine of Gautama, but it has no connection with Calvary. 'Even the modern developments of Buddhism,' writes Sir Charles Eliot, who understood the spirit of Buddhism better than most western writers on the subject, 'which represent the Buddha Amida as a saviour do not contain the idea that he gives up his life for his followers.' Renan was equally near the mark when he pointed out that what distinguishes Buddhism from Christianity, and in fact from practically every other known religion, is that it has its roots in '*la pensée pure.*' Its basic doctrine, the doctrine of Gautama which Whitehead chooses to describe as 'a tremendous doctrine,' is not, strictly speaking, a religion at all. It does not contain any sense of The Holy—that, which according to Dr Otto, is 'the "beyond" of the mystic . . . the kernel of religion.' It offers no consolations and no compensations; it is devoid of faith, and consequently of miracle, which is but faith-in-manifestation. Its central point is an act of mental apprehension and not an ecstasy; as such the emphasis is on Understanding, not on Mystery. The pattern that is projected is cerebral rather than emotional.

Gautama and Jesus stand for two distinct modes of apprehending reality. The resultant world-views have nothing or little in common. They can be reconciled to one another only at the cost of distorting both. For Jesus, Love and the *élan* of Love—an *élan*, let it be recognized, which, though transcendental to a degree, in the end involved an almost physiological rapture, a tingling of the corpuscles of the blood, a nervous transport electrically kindling the very cells of the being—was an absolute value. It needed no external sanction, no outside justification. It was for him the last word on the nature of reality. He was what we in India call *Bhakta*. To this ecstatic attitude, Gautama would have found it very difficult to subscribe. His was the cool, sceptical, and dispassionate disposition of a man of sense. Love, he would almost certainly have reasoned, is far from being the last word about anything. It is not sufficient to go through life; not sufficient to bring life's chaotic tendencies into some kind of an order. It leaves out far too much. Most important of all things, it leaves out its own fulfilment. For it is in the very nature of Love that it shall remain for ever frustrate; it is in the very nature of a craving that it can never be satisfied. These arguments, of course, could hardly have convinced the man of faith. For faith is love; and love is blind. To Gautama's dry, clear-cut intellectual expostulations, Jesus might very well have retorted in the words of Hamlet: 'There are more things in heaven and earth, Horatio, than are dreamt of in your philosophy.' And, for him, no doubt there were. But for him alone; and in the last resort, even for him they turned out to be very different to what he had dreamed.

There is little similarity between the dramas of these two lives. They are enacted on entirely different planes. The one is conceived on a Promethean scale, poignant with great joy and titanic despair. Like a feverish vision, it is vivid with fleeting transparencies and nightmare shadows of an agony 'darker than death or night.' The other theme moves through soft modulations, wistful but satisfying. Like a sober tone-poem, it seldom rises beyond a quiet meditative mood, at once diffuse and precise in its meaning. Perhaps, in the last analysis, both are tragic dramas, tragic in different ways. In one case we are given an act of faith, an act of love, leading to its inevitable, perhaps predestined conclusion in an abject betrayal. In the other, we feel something of the ache of what might almost be described as disembodied thought: a disembodied thought which, having explored the heights and abysses of the universe, discovers that it is like an eternal phantom-pantomime, a process of endless cycles of dissolution and evolution, of change and becoming; that in the end we must part from everything that is 'most near and dear to us'—even our precious thought.

So different are these two most significant figures that one cannot even approach them from the same angle, or apply to them the same standards of criticism. One can accept or reject Gautama purely on intellectual grounds. But one cannot hope to treat Jesus in the same dispassionate manner. One gets emotionally involved in him. One must either love him, or struggle against him. And in struggling against him one gets as deeply involved in him as in loving him; we have only to remember the convulsions of antichrists like Nietzsche and D. H. Lawrence, who, for all their fulminations against the Cross, have stigmata written over all their pages. It is, indeed, best either not to think of the strange Galilean at all; or to accept him without reserve, without demanding any explanations, in fullness of heart.

In an article in the *Hibbert Journal* two years ago, Sir Radhakrishnan observed that the wisdom of the Buddha 'has a singular attraction for free and enquiring minds of to-day.' The reason for this attraction, partly at any rate, is the desire to find in the wisdom of the Buddha a justification for their own beliefs and attitudes. It is quite a natural desire. The tendency to interpret another's experience to justify one's own is not at all reprehensible; up to a point, it is not only excusable, but perfectly legitimate, because inevitable. '*La vérité est,*' Anatole France writes, '*que l'on ne sort jamais de soi-même.*' This is profoundly true. One always sees, cannot help seeing, reality from the prison-window of one's own experience. No matter how detached, how objective, how dispassionate we may try to be the bars through which we look out upon the world of transient things are always present, casting their grotesquely enlarged shadow over our field of vision. We are so inexorably conditioned by our own experience that it is inevitable we should project our immediate anxieties and fears, doubts, and hesitations on everything with which we come into contact. Complete objectivity is an ideal impossible of achievement. But there are limits to which the process of self-projection may be carried. It is one thing to see the shadow of prison-bars across the horizon, and quite another to see nothing but the prison-bars; one thing to observe a dim reflection of one's own time in other times, and quite another to see in such times nothing but the reflection of one's own time. To disregard these limits is to land oneself in hopeless anachronisms.

Thus the fact that social conflicts have to-day assumed an overbearing importance has induced some writers to invest the age of the Buddha with a parallel social background; to see in Buddhism something very much akin to a mass-movement of revolt; and to picture Gautama as the leader of a political revolution on the one hand, and a religious reformation on the other. He has even been labelled 'the Divine Socialist' by a writer bearing the name of Sophia Egoroff. Other writers, with less charming names, but more serious purpose, have taken a cue from Karl Kautsky's interpretation of primitive Christianity, and tried to interpret primitive Buddhism in purely politico-economic terms. This misplaced emphasis seems to obscure rather than clarify the meaning of primitive Buddhism in general and the Buddhism of Gautama in particular.

M. Jean Przyluski, in a monograph on Buddhism, for instance, comes to a conclusion which is original rather than justifiable. '*Le Bouddhisme . . .*' he maintains, '*est a l'origine, une doctrine qui s'appuie sur les classes inférieures de la société Indienne . . . à cet égard il semble dirigé contre le système des castes.*' But, as a matter of fact, there is no evidence to show that, originally, Buddhism was a proletarian movement, aiming at the emancipation of the poor and down-trodden strata of Indian society. To attribute to it any such laudable motives is to give it credit which it hardly deserves. M. Przyluski himself seems to feel that his position is untenable; for in the very next paragraph, he admits: '*Le Bouddhisme n'est pas dirigé contre les castes en sens étroit. Il ne prétende pas davantage abolir les grandes classes sociales, ni mettre le sujet au niveau du souverain . . .*' It is reasonable to enquire that if it is not directed against the caste system in any direct and overt sense, if it does not attempt to abolish the major social divisions, and if further it does not contemplate raising the subject to the level of the Sovereign—in what earthly sense then is Buddhism revolutionary?

M. Przyluski has no satisfactory answer; perhaps there can be no satisfactory answer so long as we restrict the word 'revolutionary' to its modern politico-economic sense. Gautama was not a revolutionary in our specialized meaning of the word; at that particular stage of social organization, it would have been nothing short of a miracle had a revolutionary of this type suddenly appeared on the scene. It would be a complete falsification of facts to proclaim him as an early avatar of Lenin, or even Luther. It is unfortunate that he did not take the line of a revolutionary; had he done so, he might have emerged a more heroic and praiseworthy figure. But evidence on this point leaves little room for such illusions.

However, the fact that Gautama was not a political revolutionary, that the movement he founded had no overt political aims, does not mean that he was simply a 'holy man' who saw the problem as one mainly pivoted on mysticism and religion. August Thalheimer's contention that Gautama sought a solution within the framework of 'religious truths' only proves how difficult it is for even the best of our dialecticians to see anything objectively.

Gautama not only sought no solution within the boundaries of religious dogma, or Mystery, but was probably one of the first men of whom we know, who really saw through religious phenomena and concluded that they afforded no solution. Certainly we do not find in him any violent recoil from religion. He is not the prototype of Marlowe's Faustus. His was not the hopeless, emotional revolt of one who is so inextricably wedded to the religious myth that he would go and sell his soul to the devil to experience the presence of God in its most diabolical and absolute form—as an everlasting absence; who is so attached to the idea of the Good that he deliberately practises evil to assert the existence of its opposite; or who is willing to court eternal damnation to demonstrate to himself the reality of eternal Grace. To deny God one indeed must love him passionately, ecstatically. Gautama's attitude on all these issues was, on the other hand, that of a truly philosophic and scientific spirit; one who is 'above the battle' in the proper sense of the phrase. 'Therefore, O Ananda,' we hear him saying, 'be ye lamps unto yourselves. Be ye refuge to yourselves. Take yourselves to no external refuge. Hold fast to the truth as a lamp.' This cannot by any stretch of imagination be mistaken for the stern voice of religious authority: no religion could foster so open and honest an attitude of enquiry without seriously endangering its own foundations. Still less is this the tortured but adoring litany of a disciple of the Son of Morning. It reflects merely the serene outlook of a reasonable but somewhat unemotional person who has come to regard the paraphernalia of religious Mystery as irrelevancy rather than falsehood—as something unnecessary.

For Gautama, the only way of resolving life's endless tangles and complications lay through understanding. It was a question of adjustment of mental habits and ways of thought; a question, in the widest interpretation, of psychology. Thus his objection to the Brahmanical tradition based itself neither on moral nor on political grounds. He did not take exception to it because it tended to set up the Brahmanical oligarchy as an absolute dictatorship. It is, indeed, open to question whether at any time in Indian history Brahmans have enjoyed the supremacy so often attributed to them. At least, it is certain that at the time of the Buddha, when the caste system was still in the process of formation, they could not have done so. True, they exercised arbitration and prerogative in matters concerning the other world; and they could probably manipulate heavenly wrath at their will. But the temporal power resided entirely with the second caste in the Hindu social tetrachotomy; no matter what the Vedas said, the merchants and nobles could pay the piper and call the tune. As in feudal Europe so in classical India, the priests, while they might conjure horrifying visions of fire and brimstone for the lesser fry, maintained an attitude of obsequious indulgence towards the rich and powerful, and indeed were quite reconciled to the idea of picking the juicy bones and crumbs from the lordly tables.

Thus, if Gautama had wished to lead a political revolution, it would have been altogether more meet to have incited the people against his own caste. But he did not entertain any such intentions. His quarrel with the existing social order was not that it was unjust, but that it was stupid, and unjust because of its stupidity. He objected to Brahmans not because they were tyrants but because they were imbeciles; not because they were showing the wrong path to heaven but because they were raising hopes which could never be fulfilled, since there was no such thing as heaven. He saw in life a conflict, not between good and evil, between caste and caste, between rich and poor, but between ignorance and knowledge. Man's viciousness, his callousness to others, and in the end to himself, his failures and frustrations, he traced back not to a Fall from Grace at some remote date in antiquity but to a much more immediate cause—his failure to think properly, his failure, in fact, to think at all. Jesus said of those who struck him: 'they know not what they do.' In the context in which this was uttered, it might have meant a number of things. But taken literally it comes nearer to Gautama's point of view than any other pronouncement of Jesus.

'They know not what they do.' What is more, they *think* not of what they do. That sums up Gautama's criticism of the Brahmans as well as the non-Brahmans of his time—and of all times. As he saw it, the chief task therefore was to impart knowledge to people, to make them think. For if they knew, if they thought, their acts would be different. His was an attempt at a psychological revolution, if we can conceive of such a thing; not a change of heart only—he was inclined to be sceptical about the value and even possibility of a change of heart, and seems to have been one of the few men who have ever cared to give that overrated organ a 'socratic treatment'—but a re-orientation of men's mental horizon and habits of thought. Rightly or wrongly, he believed this to be the only change worth attempting. Rightly or wrongly, he believed, too, that if it were achieved the rest would follow naturally, inevitably.

How far was this belief justified? How far was it an evasion and wish-fulfilment—an 'ideal and fantastic victory' over a world otherwise too difficult to cope with? These are questions upon which it would be better to reserve judgement in the present state of our knowledge. So far Gautama's method has not been given a fair trial; perhaps, by its very nature, it cannot be given a fair trial. But it does present a choice of the most crucial importance; a choice not between moral right and wrong, but between truth and error. It was Goethe who remarked that the fascination of error derives from the almost complete absence of checks from its vast domain, and that what makes human beings fight shy of truth is the limitation it imposes. But perhaps one ought to add that error also has its limitations—limitations far more hopeless than any set by truth; and that, though the sphere of error be vast, its vastness is that of a huge cul-de-sac, and not of sunlit skies. The distinction between truth and error is hard to perceive; the light we possess is always uncertain and wavering, and the darkness immense. Yet the choice is there; and sooner or later mankind will have to face it. And with that choice is linked up the larger issue of humanism; of our belief in human standards, which, though not absolute in the sense in which some of the agelong aberrations of human fancy have been regarded as absolute, represent a residual wisdom, and as such not only lend dignity and meaning to human effort, but form a measure of worth with a more than personal sanction.

Humanism is not a new issue. For over two thousand years sensitive individuals seem to have been aware of the problem, or have been struggling to define it for themselves. There are several methods of approach, but finally they fall into two broad categories. On the one hand, the romantic approach—romantic, that is, not in any derogatory sense, but simply in the sense that it seeks abolition of all limitations which beset human life. Its most outspoken expression is to be found in Rousseau who said: '*J'adore la liberté; j'abhorre la gêne, la peine, l'assujettissement, l'injustice.*' A subtler, more tragic manifestation of this attitude is seen in Marlowe; for at root, Faustus' Promethean craving for 'knowledge infinite' and his hopeless attempt to capture immortality in a single kiss, represents a humanistic struggle. But the way of a Rousseau or a Marlowe, though no doubt heroic to a degree, raises more difficulties than it solves. Clearly, it is not simply a question of being passionately attached to the idea of liberty, and rebelling against tyranny. As far as freedom goes Rousseau was by no means original in his fervent longings: we all yearn for freedom infinite, and would like to be absolved of the coils which hamper us in every direction. But what we all seem unwilling to admit is that a certain amount of bondage is implicit in the very condition of our existence, is inseparable from it. For instance, although we may burn down all the Bastilles in the world, put every tyrant against the wall, and stamp out all the counter-revolutionary spiders upon whom we can lay hands, death is probably still inevitable. There is death for the tyrant and the tyrannized, for Hitlers as well as Stalins, for Baldwins no less than for Lenins. Death: it is almost a comforting thought.

But there is also the quest of knowledge infinite, of consummation in the ecstasy of love. It is a brave and noble quest. Brave and noble—but, unhappily, foredoomed to failure. Take 'knowledge infinite.' In the nature of things, alas! no such condition can exist. All knowledge to which we have access is, by definition, finite and limited; one of the principal functions of awareness is to demarcate boundaries. True, these boundaries do not remain always the same; they are ever changing; yet some kind of boundaries still remain. To know a thing is also to be conscious of the deficiency of one's knowledge of it, which is always approximate, never final. Knowledge infinite, then, is a mental chimera, a will-o'-the-wisp beyond all human grasp. So, too, the love infinite. The pity of a passion so exhausting and reckless as that of Faustus is that it demands far too much of the object of the heart's desire. Demonstrably, Helen even if she were to grant us far greater favours than a niggardly kiss, could not be expected to transubstantiate our mortal stuff into immortality. Death is no respecter of persons: Helen too is mortal. 'Whereas anything whatever born, brought into being, and organized contains within itself the inherent necessity of dissolution,' so in time all the Helens of the world once 'fairer than the evening air clad in the beauty of a thousand stars,' must grow old and weary, and slowly be devoured by the germ of decay. Such is the order of all component things.

Humanism, therefore, so long as it is involved in the mere assertion of human will seems bound to lead to a tragic culmination. Human will, whether collective or individual, cannot accomplish the impossible, cannot bring us the moon of unattainable freedom, and cannot make it possible for us to satisfy all our desires. Wisdom perhaps consists in making an effort to understand the true nature and extent of our bondage, of our deficit. For truly to understand our chains might lead us to the discovery that some at least of them are imaginary, and there are a few others which we can shake off if we work diligently.

So we have the way of Gautama. Less heroic, less spectacular by far, it possesses at least one clear advantage: it spares us the painful switchback experience of high hopes doomed to be disappointed. It makes no desperate gesture to transcend all limits; rather it is content to discover such limits as there are, and to accept them. It is not the way of Will. It does not yield to the inebriation caused by the fleeting moods of the human will in its moments of uncertain triumphs, but is aware of the abysses of dejection that open before us at every step. It sees, in brief, not only the significance of man, but his insignificance. It thus offers a humanism which is not incompatible with reason, discretion, and humility; it contains also a truly human Norm.

But there is a chorus of disapproving voices. The Buddha, if he has had crowds of admiring votaries, has not lacked critics. A typical instance comes from the New World: the case of Ralph Waldo Emerson, a direct descendant of the energetic and industrious Pilgrim Fathers. 'Always motor-minded,' his sympathetic biographer, Frederick Ives Carpenter, writes, 'he hated quietism. And as he associated quietism especially with Buddhism, he often criticized that religion. We have already noticed that he described Buddhism as being literally the tenet of "Fate." ' Ralph Waldo was, indeed, so terribly motor-minded, so anxious to step on the accelerator as it were, that he would not even pause to make sure of his facts, and talked the most disarming nonsense about Buddhism that has ever been heard on either side of the Atlantic. 'He called the *Bhagavat Gita,*' Mr Carpenter himself adds for our information, ' "the much-renowned book of Buddhism" and so showed his ignorance.' To this comment let us add, that he confessed his ignorance no less blatantly when he confused Buddhism with fatalism; for the Doctrine of Gautama, in so far as it is a doctrine at all, is as far from being a tenet of 'Fate' as the *Bhagavat Gita* from being 'the much-renowned book of Buddhism.' Moreover, if the whole argument against Gautama bases itself on his disapproval of loud-speaking, self-proclaiming, and hysterically go-getting philosophies, then there is something to be said for his quietism.

The Rev. Bruce, in his Gifford Lectures delivered towards the end of the last century, however, levels a more serious and even more typical charge against the Tathagata. Not satisfied with dismissing his doctrine as 'fantastic and morbid,' the learned divine goes on disdainfully to say: 'The well-being of the race demanded warriors brave in the field of battle against evil, not monks immured in cloister and passing their time in poverty, wearing the yellow robe of a mendicant order.' This gentleman's contempt for the Buddhist attitude should not surprise us. A view of life which does not visualize the world as one great Armageddon, naturally takes away all vestige of justification from the big battalions of Christian soldiers who have proved their mettle worthily in many much wider arenas than the narrow range of the spiritual-symbolic battleground. The impecunious, cloistral life-mode of the Buddhist monks could not possibly have commended itself to this reverend gentleman who, evidently, believed in deriving the fullest secular benefits from the revenues of a flourishing State Church. But one may justifiably wonder if a doctrine, the efficacy of which entirely depends on the existence of evil in the world, and a faith whose very *raison d'être* is bound up with the consciousness of the all-pervading nature of sin, is not more obviously open to the charge of 'morbidity' than a doctrine which, whatever its drawbacks and shortcomings, at least makes an attempt to resolve human problems by reference to an intelligible causal chain, and is not based on an obsession of the dreadful, everlasting combat between good and evil.

A more sophisticated objection to Gautama's doctrine is to be found in Mr Paul Elmer More. Until recently an eminently catholic and stimulating writer, Prof. More seems to have decided to recant his early agnostic indiscretions. He has come to the conclusion that there is no consummation to be achieved outside the embrace of a 'personal deity'; and as a result, he has thrown Aristotle overboard and joined the already teeming ranks of Lord Lorders. And like most contrite souls, ever since his appearance in sackcloth, he has turned his battery of recrimination against the company which he cultivated only yesterday. He has some unkind words to say even of the Buddha, whom in a recent essay he has compared to 'a man who has raised a stately flight of stairs that leads nowhere.'

This is perhaps poetic justice; Gautama had used a very similar argument against the Brahmans of his time. However, Gautama had a different context: he had argued that if you lead people up a flight of stairs then you must know where you are leading them. As far as he himself was concerned he was content to remain on the ground level. This takes away the point from Prof. More's analogy. Professor More is so anxious to be led somewhere that he does not mind whether he is landed into the cloud cuckoo-land of theological Heaven or down into the dream torture-chamber of ecclesiastical Inferno. Lead thou me on—*n'importe où dehors du monde*. It is his own choice.

The foregoing arguments are indicative of the main tenor of Gautama's disparaging critics. Drawn from many quarters, they include in their fold people holding the most divergent opinions, rankest materialists as well as exultant idealists, ferocious theists and even more ferocious atheists. But all have this much in common. They subscribe to positive beliefs of some sort; they affirm, each in his own way, a staunch and unshakable faith either in this world, or the next, or sometimes as it happens, wisely, in both; and they share with one another that most horrible of all horrors to which man is subject—the *horror nihili*. It certainly seems strange that these Titans of faith, armed as they are with all the impregnable virtues of a self-righteous positivism, should see in what they denounce as 'the ineffectual pessimism of the Buddhists,' a serious menace to their firm convictions, and think it fit to lash at it with an earnest fury worthy of the frantic Knight of La Mancha. It is, in fact, revealing. It reveals the weakness inherent in all uncritical positivism, whether of the idealistic or materialistic brand. What Rochester so aptly called 'balloons of philosophy,' no matter of what variety and colour, are all liable to burst at first contact with the actualities of human experience. It seems, however, that the balloon of *positive* philosophy is made of still more flimsy substance. It is not even a proper balloon, perhaps: it is more of a bubble.

It would serve no purpose to enter into the fruitless though popular controversy as to whether Gautama was a nihilist, and if so, to what extent are we justified in regarding his doctrine as of a purely negative character. But this much must be emphasized: the misapprehensions on this point arise chiefly from certain incorrigible mental fixations which have been passed on to us by the 'romantic Indianists' like the Schlegels who, on their own admission, went to Indian thought in search of 'the supremely romantic.' And to them, the supremely romantic definitely meant the supremely desperate, as also the supremely despondent. Not for nothing had they read the Sorrows of Young Werther; not for nothing had they imbibed the education of *La Nouvelle Héloïse*. De Musset was voicing a sentiment more characteristically Germanic than Latin when he wrote:

> *Les plus désespérés sont les chants les plus beaux*
> *Et j'en sais d'immortels qui sont de purs sanglots.*

It was under the influence of Frederick Schlegel's *The Language and Wisdom of the Hindus* that Schopenhauer was first drawn to Buddhism, and later came to look upon the Buddha as so much of a kindred spirit that he is known to have kept the image of Sakya Muni in his bedroom. And ever since, in many western minds, Gautama has been associated with Schopenhauer. We do not know if this association has been instrumental in bringing us better appreciation of the German philosopher, but as far as the Indian sage is concerned, it has proved extremely unfortunate. It is this misalliance which is principally responsible for our twisting Gautama, one of the most lively as well as sympathetic of men, into the rôle of a lachrymose and embittered misanthrope. It is true that there is a certain measure of parallelism between the two world-views at the outset; both starting with the recognition of the tragic element in life. But there is a very important qualification to be made. In one case the anguish begins and ends within the confines of an egotistic personality, compensation for this being found in an assertive Will that is, ultimately, self-destructive; in the other, the sense of grief leads to a realization of the all-embracing problem of suffering, thus releasing unsuspected springs of compassion. There is here a fundamental contrast: the contrast between an adolescent and adult view of life.

Whatever the nature of Buddha's 'nihilism,' and whether it is pessimistic or not, one thing is certain: it does not need apologies. To apologize effectively for Gautama would require an understanding equal to his; and there is no reason to believe that either his admirers, or his critics, possess that degree of comprehension and comprehensiveness. On the contrary, there could be no more convincing proof of the complete bankruptcy of our spiritual criteria than the fact that we know people not by the intrinsic worth of their deeds, but by the slogans they shout, the banners they wave; and that we judge an attitude not by the vitality of experience which sanctions it, but by the arbitrary labels attached to it.

In the year 819, the Tang Emperor, Hsien Tsung, came to hear that the Fa-men monastery in Shen-si housed a bone of the Buddha which was known to exhibit miraculous powers every thirty years. As these miraculous manifestations were expected in that very year, the Emperor gave instructions that the prodigious relic should be brought to the capital, kept for a while in the Imperial Palace, and then sent on a circular tour of the monasteries in the city. This was more than Han-Yü, a statesman and writer of unusual gifts, could tolerate. He worded a petition to the Emperor denouncing the Buddha and Buddhism, and protesting against the policy of allowing an outlandish and barbaric faith to contaminate the ancient civilization of the Middle Kingdom. 'Your Servant is well aware,' wrote the Celestial, at once indignant and apprehensive, 'that your Majesty does not do this in the vain hope of deriving advantage therefrom, but that in the fullness of our present plenty there is a desire to comply with the wishes of the people in the celebration at the capital of this delusive mummery . . . For Buddha was a barbarian. His language was not the language of China. His clothes were of an alien cut. He did not utter the maxims of our ancient rulers nor conform to the customs which they handed down. He did not appreciate the bond between prince and minister, the tie between father and son . . .'

Buddha was a barbarian! History has a way of emphasizing itself in repetition. Han-Yü's bitter complaint is echoed by a modern British academician. Professor Berriedale Keith, who manages to achieve in his many-sided personality a happy combination of expert knowledge of British constitutional history and an equally impressive erudition in Sanskrit literature, does not adopt the Chinese patriot's straightforward manner of invective. Both by training and preference, he is inclined to speak in a subtle and oblique fashion. After suggesting that the Buddha was the child of a barbarous or semi-barbarous age, he leaves us to draw the all too obvious conclusion. He has some particularly scathing remarks to make about the western criticism, which, 'ruthless in probing the claims of its own sacred scriptures, has treated the Pali canon with a respect so profound as to regard with open hostility any attempt to apply to these sources of information the same dispassionate scrutiny which is demanded from the researches into the history of Christianity.'

The justification for this sound and timely corrective is not for us to question. On all Indian matters, one feels strongly, the judgement of our accepted professional interpreters has always been warped by an ignorance which manifests itself either in the form of an excessive enthusiasm, or an ill-concealed prejudice. From every point of view it would be eminently desirable if the treatment of Indian subjects, from religion and philosophy down to cookery and beauty-culture, were informed by a more critical spirit than has hitherto been in evidence. We might further grant Prof. Keith that it is not possible that early in Buddhist thought we encounter 'fully appreciated ideas which have slowly and laboriously been elaborated in Europe and are normally regarded as the particular achievement of modern philosophy.' But when he goes on to say that, given the psychological conditions of his times, it would have been a miracle had Buddha been capable of the rationalism imputed to him, he stretches the argument to make a point which it does not seem reasonably to warrant.

This is historical determinism served with a vengeance, such as one would not expect from the most ardent novitiate of the Marx-Engels Institute. If its implications were accepted, one could similarly throw doubts on the 'wisdom' which it is customary to attribute to Socrates; for the psychological (and even the politico-economic) conditions under which he lived were not so very different to those experienced by Gautama. If the rise of Buddha was contemporaneous with the triumph of the great gods *Siva* and *Vishnu*, at the time of the enactment of Socratic drama, the glorious days of the Olympian patriarchal family—of Zeus and Hera, and the vast number of their progeny born both within and without the confines of the originally incestuous wedlock—were by no means over. Socrates himself, if the version given in the Apology is to be believed, received instructions of his life's mission from the god of Delphi. And even while the Sun of Hellenism was approaching its zenith, the popular soul of Greece continued to luxuriate in its deep slumber, enveloped by the soft and soothing mist of what Dr Preuss appropriately designates as 'primal stupidity' (*Urdummheit*).

The Eleusinian Mysteries, with their promise of magic and miracle, their richly entertaining sacramentalism and still more entrancing demonology, claimed an incomparably larger number of votaries than all the academies of the Eleatics put together. Both Zeno and Plato reserve bitter comment for the esoteric cults and secret doctrines that were in vogue in their days, and to which the Pythagoreans had added the alluring oriental fantasy of metempsychosis (Marion Bloom's 'something with hoses'); and from this it may be inferred that their appeal to the mass mind was far more potent than that of the sober, philosophic reflections of the early gnostics, one of whom at least was made to drink the cup of hemlock for the pains he took to teach his fellow citizens 'to care for their souls'—an example of intolerance for which, incidentally, Gautama's India, despite all its 'barbarism,' has no recorded parallel. The old Bacchic wine in the course of its transference into new Orphic bottles had lost none of its delightful orgiastic flavour: the superstitious ritualism, which was the legacy of Greece's 'dark ages,' continued to flourish in an unbroken tradition till it was absorbed almost in entirety by the Mystery of the newly-born Christian Church. Human sacrifice had more or less disappeared, but Sparta had discovered a satisfying substitute for it in the gruesome, but enthralling, practice of flagellating tender youths as a means of inducing spiritual frenzy. Such was the glory that was Greece.

We have been too much in the habit of seeing the Greek view of life through monumental marbles. These dignified museum exhibits tell only a part of the story; and perhaps there is another side which is not quite so edifying. Even the great luminaries of the Periclean age seemed less inspiring to their own contemporaries than they appear to us. Plato may call Socrates 'the wisest, the justest, and the best of all men'; but to Aristophanes he appeared a comic figure swearing by 'Chaos, Respiration, and air in his Thinking-Shop.' Pericles considered purely as a lover of the muses is a sympathetic person; but as the lover of Aspasia he was an object of ridicule to the Athenians who, we are told, 'had occasion to execrate the memory of a man who by his example corrupted the purity and innocence of their morals, and who made licentiousness respectable, and the indulgence of every impure desire the qualification of the soldier as well as the senator.' Sophocles' authorship of 120 tragedies points to a more than Shakespearian industry; but his private life, according to Athenaeus, was far less circumspect than that of the Bard of Avon. The somewhat later figure of Diogenes walking through the streets of Athens with a tub on his head, telling Alexander to get out of his light, is a supreme example of philosophic detachment; but the Cynic of Sinope, before his apprenticeship to Antisthenes, was known to have been banished from his country for coining false money. All these are no doubt minor matters when we consider what we owe to these eminent Greeks, but it is well to remember them when acknowledging the legacy of Greece.

It is equally important not to forget that the pursuit of the arts, letters, and sciences by a small minority was rendered possible only because of the slavery which formed the basis of Athenian prosperity. Lost in a loving contemplation of the delicately moulded curves purposefully concealed under the flowing draperies of the Caryatids of the Erechtheum, or pondering the voluptuous possibilities revealed by Alcamenes' Aphrodite, or rather its Roman copy, our elderly professorate of 'Humanities' is a little apt to forget that the balanced repose, the harmony, the ideality, the naturalism, and all the other 'lamps' of the art which transports them into a state of such pulsating excitement, rested on a broad base of human misery.

Indeed, if one wanted to condemn Socrates merely on the score of the moral depravity and the intellectual confusion which characterized the milieu in which he lived, it would be possible to make a much more devastating case than could be made against Gautama on similar grounds. But to reason in this manner would be manifestly unfair. The 'miracle' that Prof. Keith does not regard as possible has been happening throughout the ages. Sanity has been known to exist in the midst of rampant lunacy; wisdom has flowered again and again despite ubiquitous human folly. This merciful contradiction runs through the whole of human history; and to it we owe the small measure of civilization which has entered into human intercourse. After all, the world in which we live to-day—a world which periodically indulges in blood-baths, and subjects vast populations to the most terrible privations in the midst of unparalleled plenty—has not much to distinguish it from a madhouse. Yet it would be taking an unnecessarily pessimistic view to suggest that our age is completely devoid of reason and normality; side by side with our sabre-rattlers and Shylocks there are still men like Professor Keith.

Dean Inge says 'the spirit of man does not live only on tradition; it can draw direct from the fountain-head.' What he says is true to this extent, that there is a residue of human sanity from which gifted individuals can hope to draw wisdom at all times. And although it is true that Gautama did not have the advantage of being conversant with 'the conceptions of rationalism, of psychology without a soul, of Kant, of Schopenhauer, Von Hartmann, Bertrand Russell, Bergson *et hoc genus omne*'—an advantage which Professor Keith undoubtedly possesses—it remains nevertheless possible that, working on his own experience, he arrived at an equally reasonable view of life.

Yet Professor Keith's argument raises a problem of considerable importance. How far can we regard Gautama as a typical product of his historical background, and how far can he be said to stand outside it? Goethe suggests that the excellencies of an individual are his own; his defects, of the age. In suggesting this arbitrary method he was merely echoing a current fallacy which gave to the individual an altogether undue importance in the scheme of things. No individual, however great, is entirely above his age. But, on the other hand, no truly gifted person is merely a symptom of his times. We may, up to a point, agree with Lytton Strachey that human beings have 'a value which is independent of any temporal processes.' The crucial test of genius is not only the extent to which it participates in the peculiar idiosyncrasies of the soil upon which it grows, but also, and perhaps even more, the extent to which it rises above them and expresses itself in a new and individual form. Seeing Gautama in this light, we find that, while he doubtless shared the virtues and failings of his environment, he had also crystallized in his person a certain synthesis which was of his own individuality. He lived fully the life of his own period, but he also went far beyond it. And the quality of vision which is needed to see beyond one's times, is even more significant than the quality needed to observe the minutiae of one's immediate horizon. For while it is possible to point to a number of characteristically 'period' figures among Gautama's contemporaries, hardly any one of them leaves the impression of being in possession of anything approaching his universality. As such, in considering him, we must not only consider him as a part of his background, but also as apart from it.

EPILOGUE

IL *y avait une fois La Réalité*, says Aragon cryptically. This applies to the Buddha, too: there probably was some reality before the mythographers got busy weaving the intricate webs of their fancy. Divested of these layers of excrescence, the kernel that remains seems to be a profoundly human fact. In passing I have attempted to place this fact in some kind of an intelligible relation to other facts of a similar order, both in the past and the present. The purpose of my argument has been to suggest not only that in Gautama we have a very sensitive individual, a representative figure who sums up in himself the effort of a distant epoch to synthesize the ecstasy and delirium of a nascent self-consciousness, but that in him we have an example of truly humane and civilized sensibility; and I find myself in complete agreement with Irving Babbitt when he expresses the view that the urbanity resulting from Gautama's influence in India was perhaps the nearest approach that that very unhumanistic land has made to humanism.

The synthesis which Gautama realized in his person is at once a disturbing and saddening example. It is disturbing not because I believe that Gautama belongs to a species of super-men for whom Schweitzer has invented the hyperbolic concept of 'Immeasurably Great.' Fortunately perhaps, 'Immeasurably Great' men do not exist outside the enchanted world of fairy tale and day-dream. If Gautama, like Schweitzer's Jesus, were an Immeasurably Great man, then he would be of little value to us. The effective worth of a man, even of a hero among men, can be determined only in terms of his humanity. And it is precisely this essential humanity of the man that is so disturbing about Gautama. He points to a quality of awareness, a way of life, which is not at all an abstraction, but appears to be something tangibly within human reach. And yet for some obscure reason it has so far eluded the grasp of men. They have either been afraid of it, or they have rejected it without caring to examine its meaning; and where they have tried to realize it, their first act seems to have been to distort its vision of light and order into a nightmare of confusion. That is the sadness, the unutterable pity of it.

The nature of this Way, as also the process through which it was reached, I have tried to set forth—at least as a tentative outline. In order not to create a wrong impression, it is perhaps necessary once again to say that I do not look upon Gautama's Way as an *Open Sesame* to the Land of Heart's Desire. Unhappily, there is no such magical formula; for there is no such Land. What Gautama offers one is merely a way of understanding. Therefore, those who are after talismanic effects are likely to be left with a keen sense of disappointment in the last analysis. For it seems to be a way, which, contrary to the current belief, does not lead to the ideal realms of endless bliss: it only brings one to an observation post whence one can take a broader, more comprehensive, and calmer view of the outstretching expanse of life's desert territory, on the fringe of which mirage after mirage rises in quick succession, dances its brief turn, and disappears. Gautama does not pretend to heal the lepers: he merely says that lepers too can, if they make the effort, see their sores for what they are, probe to the root of their disease.

I confess that this is hardly a comforting ministration for an afflicted, long-suffering humanity. But in the long run it seems to me that the real remedy, if there is one at all, lies in discovering the cause rather than fostering hopes of imaginary cures. 'It is the cause, it is the cause, my soul,' cries Othello. It is with the cause that Gautama, too, concerns himself. He does not deal in anodynes: he merely affirms that in the measure in which pain understands its own nature, and ceases to be the blind agony of an embryo —in that measure it also surpasses itself.

It would be romantic to suggest that his is the last word. No word represents finality: it is always the shadow of a thing and not the thing itself. But in the antelucan world of shadows, through which the spirit of man has all this time been trying to feel its way to light, it appears to come as near grasping the sense as well as the mystery of things as any other word uttered before or since.

LITTLE TO ADD

SINCE the last chapter was written a great deal has happened in the world at large. We have been living through a turbulent age of wars, revolutions, and counter-revolutions, which, for good or ill, have significantly altered the whole context of the debate as to how and where the Buddha connects with humanism, and many of the issues discussed earlier will doubtless need to be re-examined—if for no other reason than the fact that these secular developments have profoundly affected the fate of some of the countries where Buddhism had its main historical base, with large sections of their populations professing, at least in name, abiding allegiance to the Buddha. Such a re-examination, however, does not strictly fall within the scope of this book. What is, perhaps, relevant to point out is that the Second World War inhibited the growth of Buddhist studies both in the East and the West; and during the war years, and for some time after, the flow of works on the Buddha and Buddhism in general was reduced to a mere trickle even if it did not altogether cease.

The flow has since been resumed and has been slowly but steadily increasing. Indeed, in the early 'sixties we witnessed something of a spate in publications because of an important date in the Buddhist calendar—the 2,500th anniversary of the advent of the Buddhist era. How precisely this chronology was worked out remains a little obscure. But, then, this is a matter on which wide divergencies have always persisted and are likely to persist. Thus Dr G. P. Malalasekera, who was at the time President of the World Fellowship of Buddhists, in his booklet *The Buddha And His Teachings*, prepared in 1957 for the Buddhist Council of Ceylon 'as part of the Buddha Jayanti Birthday activities', opted for 623 BC as the year of Gautama's birth which, on the available historical evidence, would appear to be a rather eccentric and unlikely guess.

Be that as it may, the occasion was celebrated with appropriate pomp and circumstance under the patronage of ruling establishments in many countries, among them Ceylon, Burma, Thailand, and, of course, India where in recent years socio-political rather than philosophical or spiritual reasons have led to the emergence of a neo-Buddhist movement among the lower castes, rightly impatient to secure for themselves equality of human rights which in theory the Indian Constitution and laws guarantee them, but which entrenched prejudice and custom, reinforced by a morbid religious bigotry, too often deny them in practice.

Surprisingly, even Maoist China participated in this ritual of adulatory pieties and intoned verbal tributes to the World-Renouncer of Kapilavastu, though obliquely and, as it were, *sotto voce*. Surprisingly, because consistent with its particular, not to say selective interpretation of Marxism, it had not always taken kindly to the Buddha or his devotees, having earlier classified the Buddhist monks in the good company of Taoist priests, fortune-tellers, and soothsayers as parasites, and deprived their temples and monasteries of much of their land as part of the Land Reform movement. What is more, as early as 1953 at a command performance staged in the Kuang-Chi temple in Peking the Buddhist Sangha (Brotherhood) of China, such as it was, was reincarnated as the Chinese Buddhist Association under the presidency of the venerable monk Yuan-Ying who died later that year. Hostile critics of the regime have been inclined to look upon this Association as a departmental extension and instrument of the temporal power.

And well it may be. But in fairness it must be added that if Chairman Mao thought it fit to domesticate the Buddhist institutions in his vast domains to suit the purpose of the state he had founded, he could invoke respectable precedents. A similar policy was initiated by the T'ang emperors and successfully operated by them; and the Kuomintang tried to do the same but lacked the will to implement it effectively. At any rate, the rulers of New China even if they consider the Buddha's way of all-embracing compassion to be as deadly a heresy as Revisionism, with characteristic pragmatism have realized that some of the relics supposed to be associated with his person have their uses. For instance, a tooth of the Buddha claimed to have been unearthed in China—and he seems to have had exceptionally good teeth for them to have served him well till his dying day and some are apparently still in excellent state of preservation two and a half millennia later—was sent out on a quasi-diplomatic mission for public display to Rangoon in 1955 and, again, in 1961, to Ceylon or Sri Lanka on a loan exhibition, though, understandably, never to Tibet. There can be little doubt that these exhibitions of an intimate object of faith and veneration, straight from the Buddha's mouth, so to speak, helped to consolidate China's friendly relations with these and other Buddhist countries.

What must, however, remain a matter of conjecture is whether all the impressively pious and festive ceremonials connected with the observance of the 2,500th anniversary of the birth of Gautama Buddha in various parts of the world stimulated serious and adult interest in the man and his way of life, or were even instrumental in bringing about any significant breakthrough in our knowledge of the forces and pressures which determined the phenomenal growth of Buddhism as a historical–religious complex in the countries of South East Asia and the Far East while it withered away in the land of its birth, leaving behind at best a vestigial if tantalizing memory.

It would be nice to be able to claim some notable advances in either of these two fields of research; nice, but hardly accurate. On the contrary, a critical sampling of a representative cross-section of the literature dealing with the Buddha or Buddhism which has come off the printing presses during the past quarter of a century, at least in English, is apt to leave one with the pessimistic feeling that the great and golden age of Buddhist scholarship which began in the second half of the nineteenth century now lies well behind us. It ended some time in the late 1930s when, in a true sense, the nineteenth century ended, too. Since then one has been aware of a certain diminution of the scale and stature of effort in this area; a narrowing of the range and scope of insights; a blurring of the focus of vision; and—it has got to be admitted—a distinct falling away and decline of the quality of understanding.

...

This is by no means to suggest that in the specific fields, whether of descriptive history, or philosophical interpretation, or biographical exegesis, or textual compilation and translation, the past few decades have been wholly barren in notable contributions. That would be an exaggeration. Take the history of Buddhism, for instance. Nothing comparable to Sir Charles Eliot's masterly presentation in his *Hinduism and Buddhism* (three volumes) has come our way in recent years and, perhaps, is not likely to for many a day. But in his chosen sphere Kenneth K. S. Ch'en in his *Buddhism in China: A Historical Survey* (Princeton University Press, 1964) has offered us a lucid and highly readable account. If the early part of the book is at times flawed by his rather uncritical acceptance of the Buddhist legend and loose attribution of certain doctrinal beliefs to Gautama, this is more than compensated for by his admirably objective analysis of the policies of the Kuomintang and the Maoist regime towards Buddhism and the Buddha. This is the more remarkable because it was written at a time when the wave of anti-Chinese and anti-Communist hysteria was still running high in the United States, sweeping many academics off their balance and grounding their reason.

Again, Dr and Mrs Rhys Davids are no longer with us; gone, too, are Dr E. J. Thomas, Professor Irving Babbitt, M. René Grousset, Professor S. Dasgupta, Dr Paul Dahlke, Dr Har Dayal, and Dr Radhakumud Mookerji. They brought to bear on the historical and philosophical themes of Buddhism they touched a quality of perceptiveness and originality of mind which compelled the deepest attention even if they did not always secure from one a willing suspension of disbelief. It would be disingenuous to say the same of the latter-day Buddhist scholars. Yet Dr Edward Conze's studies, and especially his *Buddhism: Its Essence and Development* published in 1951 (Bruno Cassirer, Oxford), is characterized by a sincerity and gravity which must command respect even where the argument fails to carry conviction.

The late Arthur Waley, to whom so much is owed for his translations of Chinese poetry and *Introduction to the Study of Chinese Paintings*, in his preface to Dr Conze's book, wrote: 'Books are, to my mind, valueless unless they express a point of view, and they must do this not by distorting reality, but by making apparent to the reader the Author's emotional and intellectual reaction to these facts. Dr Conze's book . . . succeeds in doing this.' This is a judgement with which one can go a good part of the way. The point of view which Dr Conze adopts is, of course, a very catholic one, as would be expected from a scholar who is deeply steeped in the lush heritage of mystery and metaphysics, magic and mythology, of the Mahayana, or the Greater Vehicle, which in its tidal sweep northwards and eastwards from India took in within its warm embrace practically every religious belief and practice, and even superstition it encountered, rejecting hardly anything, no matter how incongruous with the original enlightenment. But it is a point of view whose legitimacy within its latitudinarian premise it would be churlish to deny.

On a different and more popular level the same would apply to the writings of the late Sir Christmas Humphreys, although he could not be identified with any particular school of Buddhism—not even the Greater Vehicle. As we know, he was the founder and leading publicist of the Buddhist Society, London, and combined in himself a lifelong preoccupation with criminal jurisprudence and a sustained interest in what he called 'world Buddhism'. Apart from his many other distinctions, he could claim the credit for having introduced Buddhism, if not the Buddha, to the paperback market. Admittedly, in the years between the wars one used to come across on railway bookstalls on the Continent a paperback edition of Oldenberg's *Buddha: his life; his doctrine, his order* (it may or may not have been a Tauchnitz product) in the company of such classics as Anita Loos's *Gentlemen Prefer Blondes*. But that was before the days of mass tourism and the awakening among the younger generation in the West, weary of the tawdry felicities of a consumer society, of curiosity in Eastern religions, especially those at the more esoteric end of the spiritual spectrum. In his *Buddhism* (a Pelican book published in 1951), Sir Christmas provided an easy access at once to a predigested history and the teaching of various schools of Buddhism without taking sides, though this is not to say that he did not know who is or is not within the pale.

He was to continue his work as *vulgarisateur*—in the best French sense of the term—by bringing together in one volume the *Wisdom of Buddhism* (Rider & Co., 1960), which, presumably, is not the same thing as the wisdom of the Buddha. He explained, rightly of course, that 'there is no such thing as Buddhist "Bible" in the sense of an authoritative statement of the teaching of Buddhism, and this for two reasons. First, because the Buddhists recognized no authority, human or divine, for what is truth, or even for what Gautama the Buddha taught mankind; and secondly, because the range of country, language, and time involved in the term Buddhism makes such a volume impossible.'

No doubt it does. All the same, whether or not there is such a thing as a Buddhist 'Bible', or is even conceivable, the urge to provide something in the nature of a Buddhist vademecum for the greater spiritual comfort of the faithful has always been strongly felt from age to age, not least in our own day and season. To this *genre* obviously belongs G. F. Allen's *Buddha's Words of Wisdom: The Buddhist Companion-Book* (George Allen & Unwin, 1959). It offers us '365 maxims and utterances attributed to Gautama Buddha for each day and night of the year', compiled from the Pali Canonical writings. As for the extra day and night of the leap year, one may assume, the devotees are expected to fend for themselves and find something appropriate from the vast treasury of edificatory Buddhist literature.

If the aim of the compilations of Sir Christmas Humphreys and Mr G. F. Allen is to make things easy for the faithful or those in search of faith, this can scarcely be said of the *Living Thoughts of Gotama Buddha* (Cassell & Co., 1948) presented by the late Dr Ananda Coomaraswamy in collaboration with Miss I. B. Horner. Quite the reverse. Their 'pre-text' has all the virtues of self-conscious profundity and, inevitably, the defects of those virtues too. Indeed, the end result of 38 pages of introductory and expository essay is general obfuscation rather than any particular illumination; and the 'Living

Thoughts' which are gathered for our benefit tend to be selective rather than representative. On one point, at any rate, one can agree fully with their judgement on Gautama's doctrine: 'It is a doctrine for those whose wants are few, not for those whose wants are many.'

In the field of translation, the recent period has produced but a lean harvest. Perhaps, after the magnificent effort embodied in the Buddhist section of the *Sacred Books of the East* and the publications of the Pali Text Society, there is little more that remains to be done, though it is arguable that each generation, ideally speaking, must do its own translations. Two significant English translations of Buddhist texts, however, have been published in the past three decades. The first is E. M. Hare's translation of the *Sutta-nipata* under the title *Woven Cadences of Early Buddhists* (Oxford University Press, 1945). The English title, we learn, 'is suggested translation of the two words Sutta nipata, though not in accordance with Buddhaghosa's rendering. The suggested translation is altogether felicitous. For, certainly, the translation is delicately cadenced and, what is more, manages to capture something of the engaging, if whimsical, lightness of touch which seems to have distinguished these early Buddhists even when they were preoccupied with eminently didactic themes and remembering the words of one whom, ringing the changes, they call 'the Sage', 'the lone wayfarer', 'the rapt muser', and 'the awakened one'.

As enchanting, if not even more so, is a much smaller translation into English of a rather late but rare Tibetan text by Dr Edward Conze, with a preface by Professor J. Bacot (Bruno Cassirer, Oxford, 1955) entitled the *Buddha's Law Among the Birds*. The *Bya Chose* (to give it its short Tibetan title), Professor Bacot remarks, was 'first published in Calcutta in 1904' by Satish Chandra Vidyabhushana who 'does not tell us whether his text was based on oral recitation, a

Manuscript, or a blockprint'. Dr Conze's translation, he adds, 'owes a great deal to the French translation by Henrietta Meyer, brought out in 1953 by the *Cahiers du Sud*, under the title of *Précieuse Guirlande de la loi des Oiseaux*'. But whatever his source Dr Conze's translation has compelling charm, as indeed has the story.

It opens, as Dr Conze notes in his commentary, 'with a verse from the Buddhist Scripture which reminds us that, in his infinite compassion, the Buddha has brought the doctrine not only to men, but to all living beings. It then goes on to tell us how the Dharma reached the birds, through the medium of a Bodhisattva, disguised as a cuckoo . . .' It might well be asked: why a cuckoo, that most ambiguous of birds? The question, of course, would appear superfluous to the Buddhists. For, as Dr Conze rightly says, 'in the Jatakas we read stories describing how the Lord Buddha, when still a Bodhisattva, was born as a peacock, a goose, a quail or a parrot'. An added reason for the choice of a cuckoo, described here as the 'Great Bird' and even the 'Great King of Birds', could be, as Dr Conze suggests, 'the manner in which he [the cuckoo, that is] disposes of his family responsibilities'. This evidently 'endears him to the Buddhist hearts', possibly because 'Buddhists in general are apt to regard family life as stuffy and depressing', to be forsaken for the freedom of a monastic life.

What matters, however, is that the Great Bird who 'instructs and guides the religious discussion of the assembled birds' is reasonably successful in persuading them of the virtues of the Good Law. 'Now that you have achieved a certain knowledge of the Dharma, and arrived at a belief in the law of Karma', he asks them at the end, 'let each of you herewith make a promise. Exert yourselves in the bright acts of merit, and shun the dark acts of evil. Let all now firmly promise some improvement.' And they do:

The *King Vulture* then promised not to kill another living being. The *White Grouse*, the heavenly bird, promised hereafter to live only in mountain ranges, never to descend into the valley. The *Goose* promised to seek food only from the waters and the swamps. The Indian *Kestrel* promised to spend but one hour a day on food. The *Pigeon* promised to make his home in the hollow of a rock-cavern, and to respect the slopes and the shrines. The *Lark*, one of the smaller birds, promised to offer worship to the three treasures with melodious song. The *Cock*, the domestic bird, promised no more to violate the dawn. The *Hoopoe* promised to make the caves his winter home. And all the birds, both large and small, promised not to gather food above their daily needs.

Not quite all, it seems. For there were two exceptions. 'The *Raven* and the *Kite*, from habitual avarice,' we learn, 'would make no promise at all'. But the Great Bird was well satisfied and concluded with an admonition that might well serve as an appropriate text for those who are in the habit of assembling every year at Turtle Bay for the General Assembly of the United Nations: 'Henceforth let this be your Dharma! O Tibetan birds, you have prospered here in Tibet, join your songs in harmony with those of other birds.' Perhaps this somehow does connect with what Gautama thought and felt, however tenuously. Who can say?

. . .

The Buddha was never fortunate in his biographers. While many writers of varying talents, from Asvaghosa downwards, have been at great pains to magnify him beyond human proportions, few have troubled to give us a measure of his true humanity, possibly because that is an infinitely more difficult proposition. At any rate, recent years have not added much of enduring value in the field of biography, which has remained largely uncultivated for reasons which are not difficult to understand and will be briefly analysed presently.

The *Life of the Buddha* by Adam S. Beck, introduced by G. F. Maine (Collins), for instance, is modulated on a note of high piety meant to edify. But at times it reads rather as if Charles Lamb were born again to write a tale of Gautama Buddha which could be safely read in the nursery. On the other hand, the *Life and Times of Buddha*, subtitled *Portrait of Greatness*, text by Gabriele Mandel Sugana and translated by Vivian Hart (Paul Hamlyn), is a glossy technicolour production obviously aimed at the popular international audience. But here the pictorial material, the relevance of which is not always self-evident, overshadows the text, though this is perhaps not surprising in an age in which the audio-visual media are fast gaining ascendancy over the written word.

However, all has not been weariness and vexation of the spirit over the past decades. At least there has been one noteworthy biographical contribution: Professor A. Foucher's the *Life of the Buddha according to the Ancient Texts and Monuments of India.* First published in 1949 in France (*Éditions Payot*) as *La Vie du Bouddha*, an abridged translation by Simone Boas was issued by the Wesleyan University Press, Middletown, Connecticut in 1963. The merit of Professor Foucher's book is that it evokes, not so much the lineaments of a distant divinity, as the authentic portrait of a very humane man. He makes the very valid point—a point which is, indeed, implicit in our whole argument—that 'too often along the way [in his explorations, that is] we have perceived the beating of a heart beneath the impassive bearing of a pretended god; too many typical anecdotes have been recounted, too many meaningful words have been recorded, credited to the Blessed One, not to leave an impression of his personality.' This is very true.

What is no less true is that if the heartbeat were not there, the rest would have little meaning for humankind. What Professor Foucher is impressed by in that personality, as indeed Gautama's own contemporaries were impressed, whether they happened to be kings or devout courtesans, are the 'innate distinction', politeness, urbanity, even a practicality and fastidiousness which not only disapproves of 'lust, theft, lies and cheating' but also of 'carelessness in dress, the licking of fingers when eating, speaking with a full mouth, ogling the food in the bowl of one's neighbour'. He goes on to conclude with a somewhat unusual claim on behalf of the Buddha. 'We may then conclude', he observes, 'that the Buddha Sakyamuni was a gentleman to his finger-tips.'

What that means is anybody's guess, though it seems highly unlikely that he would posthumously qualify for the membership of the Carlton Club or even some of the working men's clubs which seem to be no less exclusive of strangers whose shade of pigmentation differs from their members'. Professor Foucher is, however, on firmer ground when he says that Gautama was 'free from the lightest suggestion of charlatanism and fanaticism as well as endowed with a surpassing strength of mind and perfect serenity. He also was an austere moralist but without excess and considerate of others, an independent and judicious thinker as consistent an enemy of meaningless metaphysics as of vulgar superstition.' A man, he might have added, for all seasons if ever there was one, with an imperturbable sense of outward as well as inward poise, serene, not because he was indifferent to the human condition, but because he was deeply aware of it; combining compassion with dispassionateness, as only those rare beings can who have succeeded in reaching the other side of despair.

...

Professor Foucher, unhappily, is the exception that proves the rule, although it can be said that M. André Bareau in his more recently published *En Suivant Bouddha* (Philippe

Lebaud, Paris), treads partly the same path—but with a difference. A Professor at the *College de France*, he is regarded in France as a great living authority on Buddhism, which is not quite the same thing as the Buddha. He had already published in 1985 in a collection entitled 'Philosophers of All-times', the canonical texts of Buddhism (Edition Seghers). Those texts are incorporated in his latest work, the title of which *In Following Buddha* recalls the late Professor René Grousset's famous book *In the Footsteps of the Buddha* which was published in the 1930s (an abridged version of it was published in the post-Second World War period). However, it would not be unfair to say that Professor André Bareau's book has nothing like the philosophical depth or the imaginative sweep of the late Professor René Grousset's work which may well underline the difference between the old Buddhist scholarship and the new.

In some sense it is also symptomatic of the weakening of the academic critical faculty—and not only in France which goes with excessive specialization. Professor Bareau, while selecting his Buddhist texts, makes little attempt to discriminate between those which have a higher or lower coefficient of credibility, often giving the impression that he rather prefers the texts in which the legendary and even mythological fancy is given ample play. This may be because frequently Professor Bureau opts in the choice of Buddhist canonical texts, such as the *Vinaya Pitaka* and *Dhammaguptaka*, for their Far Eastern and even Siamese rather than Pali variants. This, despite the fact that at one stage in his introductory chapter he quotes the Latin adage which means to translate is to betray—especially in a language so different as Pali and the Chinese.

This, of course, is not to suggest that Professor Bareau has not examined the Pali canonical texts of essential Buddhism. But too often his references are to the Far Eastern versions.

The result is that the Buddha that emerges from them is quite unlike Professor Foucher's Buddha—not at all an urbane and modest person, but one who tends to be vainglorious and boastful of things he has accomplished when he lived—and in his previous lives. This is all the more surprising because Professor Bareau admits that the texts he has quoted in their will—*volonté* is the term he uses in French for which their is no exact English equivalent—to achieve an 'incantatory' effect, tend to be 'clumsy' and 'naïve'.

However, Professor Bareau, in the ultimate analysis, is on the same side of the barricades (inappropriate though that metaphor is in the context) as Professor Foucher and the present writer. He sees in the Tathagata a man rather than a divinity. He says, in effect, that the silhouette of an historical personality stands out through 'the golden mist of legend' who lived six centuries before Christ and lived an itinerant life in the Gangetic basin. He further maintains that the Buddha affirmed that everything that has a beginning must have an end and that nothing 'is permanent, immutable and eternal either in beings or things'; that he denied what the Westerners call a soul ever existed. But, curiously Professor Bareau argues that Gautama believed in rebirth. He writes: 'the desire to live and enjoy leads to rebirth. This rebirth does not come about through transmigration of the soul (because its existence is denied) but through the mutation of the elements which compose the individual, whose vital principles, conditioned by the acts accomplished in the previous life, form new structures, human or animal.'

This is like contending that the Buddha wanted to have his cake and eat it too; that he wished to bring in the doctrine of transmigration of the soul, if not by the front door, by the tradesmen's entrance at the back. It is no service to the Buddha to associate him with such specious and speculative reasoning if not an intellectual confidence trick. At least for those who contend that there is 'a soul' which persists after death, belief in transmigration is a logical absurdity. But for anybody denying the existence of the soul and still maintaining that rebirth is possible through the process of desire to enjoy the pleasures of living, such as they are, is to indulge in an absurdity which is illogical.

Professor Bareau introduces another element of novel ambiguity into what is referred to in the Buddhist legend as the story of 'the Great Decease'. He suggests that the illness which led to the death of the Buddha could have been caused either by eating bad pork or mushrooms (possibly some variety of poisonous fungi). He does not quite specify the reason for this eccentric interpretation. But, surely, it misses the whole point of the story of 'the Great Decease'. What the story was meant to signify, was, firstly, that Buddhist monks, unlike monks of some other faiths, were expected to go out each day and beg for their meals instead of laying in great stores of victuals in their monasteries. Secondly, it is significant that the Buddha took his last meal in the house of a man who was an outcast, thus stressing his disregard for caste or creed. Earlier in his ministry he had refused to lay down any strict rules regarding what to eat or not to eat, presumably on the principle that those who had to beg for their daily bread could not be choosers, although his cousin, Devadatta, who tried to be more Buddhist than the Buddha, urged him to ban taking of meat or fish for those entering the Buddhist brotherhood—thus causing the earliest schism. A final point implicit in the story of the Great Decease must be that even the Tathagata is heir to the laws of mortality, including death by eating bad pork. Somehow that seems a comforting thought.

Professor Bareau in *En Suivant Bouddha*, incidentally, makes another point and it is a positive one, although it has been made before. He argues that in ancient India women enjoyed considerable freedom and were treated with due consideration. He infers this from the case of Amrapalika (or Ambapalika), the cultivated courtesan of Vaishali, who not only invited Gautama to her house for a meal, but discussed the finer points of the doctrine and even argued with him. That, at any rate, proves that the extreme misogynist views attributed to him were much exaggerated. Indeed, the latter-day Buddhist monasticism foisted on to the Buddha the misogyny which was its own perverse way of psychologically 'distancing' itself from the attraction it felt for women—an aspect of Buddhism which has been discussed in the chapter on Buddhist art.

...

As in the past so in the present, however, most writers on Buddhism and the Buddha have opted for the 'impassive bearing of the pretended god' rather than the man with the beating heart. Indeed, there are times when one wonders whether the Buddha is any longer really necessary to Buddhism. Historical experience amply proves that institutions have a life of their own; that they grow and develop according to their own logic, independently of the wishes and ideas of those who originally founded them. This is no less true of Buddhism as of other institutionalized religions. The institutions, the rituals, the disciplines, the esoteric practices, the endless ceremonials, above all the hierarchical structures which the various schools of Buddhism have evolved over the centuries would, surely, have been a source of surprise to the Tathagata had he been reborn today. He might have even been tempted to modify part of his dying words. Not decay only, but also growth and proliferation are inherent in all component things.

Not that Buddhism, at least at its popular level, is going to be able to do without the Buddha any more than Christianity is going to be able to do without Christ in the foreseeable future. But it is not being cynical to suggest that his function seems increasingly to be to supply a vastly appealing and attractive public relations image rather than the central inspiration for the structure of belief. At any rate, among some contemporary writers, it is possible to detect a distinct shift of emphasis away from Gautama's distinctive understanding of the nature of human experience and his austere thought to historical Buddhism as it has evolved over the past twenty-five centuries in all its rich and bewildering diversity. It is a fact, for instance, that in many works on the subject these days it is customary to dispose of the Buddha in a few pages and then plunge into a study of that 'family of religions and philosophies' which come under the label of Buddhism, with only perfunctory references, if at all, to their supposed source.

The reason given for this transference of emphasis is that nobody quite knows what the Buddha said or did not say. 'The Buddha himself', Sir Christmas Humphreys observed in his *Buddhism*, 'wrote nothing, and none of his teaching was written down for at least four hundred years after his death. We therefore do not know what the Buddha taught, any more than we know what Jesus taught.' Somewhat paradoxically, Sir Christmas later goes on to say: 'Gotama the Buddha was the Buddha because he was *Buddha*, the Enlightened One, and he who cannot accept this premise will never know more than the shell of Buddhism.' Maybe, but if we do not know what he taught, how are we to accept and comprehend the fact of his enlightenment? On what authority or through what process of apprehension?

Dr Conze in his *Buddhism: Its Essence and Development* argued on similar lines, though it must be said of him that he did not try to have it both ways. 'I confess', he wrote, 'I do not know what the "original gospel" of Buddhism was.' He expressed some annoyance with those who contend 'that one must seek for the true Buddhist doctrine only in what Gautama Buddha said about 500 BC'. 'To regard all latter Buddhist history', he added, 'as a record of the degeneration of an "original" gospel is like regarding an oak tree as degeneration of an acorn.'

Undoubtedly it would be in the highest degree unreasonable, not to say absurd, for anybody to dismiss *all* later Buddhist history as a record of the degeneration of the original doctrine. But reasoning by analogy is not the most convincing way of making a point, since analogies can be misleading; and the more seductive they are the more misleading they can be. Dr Conze's argument would carry greater validity if he had tried to demonstrate that later Buddhist developments are as organically related to what the Buddha said and believed as the oak tree is to the acorn. But he makes no attempt to do so. Indeed, he acknowledged earlier that 'much of what has been handed down as "Buddhism" is due not to the exercise of wisdom, but to the social conditions in which the Buddhist community existed, to the language employed, and to the science and mythology in vogue among the people who adopted it.'

This is beyond dispute. It is precisely because of this that there is some justification for regarding 'much of what has been handed down as "Buddhism" ', and accepted as such widely, as overgrowth determined by historical contingencies involved in the diffusion of Buddhist ideas across vast stretches of time and space and varieties of human culture. The process has inevitably led to the grafting upon the original doctrine, whatever it was, of disparate often totally alien elements. This is not to say that these graftings, overgrowths, compromises, and assimilations are necessarily worthless. But adulteration of the original thought there has been and it cannot be pretended that it has invariably resulted in qualitative improvement. On the contrary, it is not being unkind to say that 'the Diamond Throne of the original enlightenment [to quote Okakura Kakuzo again] is now hard indeed to discover' because of what the faithful, for reasons of their own, added to it; the intricate glosses that have been put on the word of the Buddha seem often to have distorted its meaning beyond recognition; and the light that was kindled on the 'yearning road to Benares' is all but lost in 'the labyrinths of gigantic pillars and elaborate porticoes which successive architects have erected, as each has added his portion to the edifice of faith'.

...

It is, perhaps, significant—or should one say, symptomatic?—that so eminently reasonable a scholar as Dr Conze should consider it unreasonable that 'some influential writers, bred in a Non-Conformist tradition', should contend that 'one must seek for the true Buddhist doctrine only in what Gautama Buddha actually said about BC 500'; that he should be critical of Mrs Rhys Davids for her emphasis on 'the man' as the essential element of the original doctrine; of H. J. Jennings for removing 'all references to reincarnation from the Scriptures'; and of Dr P. Dahlke for ignoring 'all the magic and mythology with which traditional Buddhism is replete' and reducing the doctrine of the Buddha 'to a quite reasonable agnostic theory'. But symptomatic of what?

The short answer is: a retreat from reason which is something more than mere rationality. It is not the first time that this has happened. One can even argue that the history of Buddhism down the centuries has been characterized by a periodic oscillation of the pendulum between two opposite trends: periods of search for its doctrinal centre of gravity in rationalism and humanism which, ineluctably, have led back to Gautama Buddha as the source of inspiration, have alternated with long periods of revelling in the esoteric cults and rituals and disciplines, with the Buddha in the role of the presiding deity of the great mystery play, if not just an impresario. In our own times, it would seem that until the late 1930s the first trend was gaining ground. Since then the trend towards mystification and even obscurantism appears to have become more pronounced if not dominant.

This is reflected in two related and mutually supporting developments: first, recourse to needlessly, even pretentiously, obfuscatory semantics; and secondly, claims that the Buddha was not trying to say anything new, but merely offered, a restatement, almost a rehash, of what Vedantic thinkers and mystics and hierophants had said before him and were to say after him. A rather naïve example of this argument was heard at the All-India Oriental Conference held in Bombay some years ago in a paper read by Professor Mohan Singh who published it under the title 'New Light on the Buddha's First Sermon, Preached at Sarnath, Benares'. Professor Mohan Singh's thesis, which has already inspired Mr A. J. Bahm, himself Professor of Philosophy, University of New Mexico, Albuquerque, to produce a book, *Philosophy of the Buddha* (Rider & Co., 1958), had best be summarized in his own inimitable words. 'By the rendition of *Arya* as noble, *dukha* as suffering, *Sam* as right,' he maintained 'the grandest sublimest transcendental teachings of the *Abhisambuddha* have been made un-Aryalike; Active *Dhayani* Buddhism has been reduced to Stoical pessimistic ethics; and the path of transcendence of the pairs of opposites, expounded in the *Gita* under Buddhiyoga, the path of balance has been made to lose itself in the medieval forest of righteousness and visionariness.' One wonders what Gautama would have made of this. But if one did not know how earnestly Professor Mohan Singh pursued his ideas, no matter how nebulous, one would be tempted to dismiss his thesis as high-falutin gibberish. For whatever was the teaching of the Buddha, it was meant to lead us in a direction almost diametrically opposite to where the Gita was meant to lead—and has led.

The same argument, of course, has had more sophisticated protagonists. In the introduction to the *Living Thoughts of Gotama the Buddha*, Dr Coomaraswamy and I.B. Horner, after stressing 'the insuperable difficulty' of conveying 'an adequate idea of the content of early Buddhist doctrine', insist that the doctrine with which Gautama identifies himself 'had been taught by his predecessors in ages past as it would be taught by his successors in ages to come'. It is not quite clear how they reach this conclusion. No doubt, like any serious and earnest thinker, Gautama had systematically studied the philosophies and doctrines, whether mystical or metaphysical, which his age had inherited and to which it had added. Indeed, he had experimented with some of the major disciplines they enjoined. But in the final analysis he had found them wanting and rejected them as 'irrelevant'. This rejection was an essential step towards his own enlightenment. His own conception of human priorities was radically different and it was meant to, and did up to a point, alter the direction of men's thought and endeavour. It was precisely in this change of direction that the intellectually and spiritually revolutionary impact of his doctrine, if it can be so described, consisted. To suggest that he was content merely to repeat what others had said before him betrays an astonishing lack of discriminating judgement and savours of deception or self-deception—and possibly both.

Whether they are aware of it or not, the argument that Gautama Buddha said nothing new and original but played yet another variation on the old Vedantic theme, has certain distinct socio-political implications. As in the past so today, it links up with and underpins obscurantism and counter-reformation—and not only in Buddhism. For while it is true that Gautama did not set about launching any specific movement of social reformation, it is no less true that his influence during his lifetime and his ideas, at least for several centuries after his death, served as the leaven for a fairly significant reformation of the Indian social structure or, at any rate, kept certain particularly reactionary and inequitable developments in check. Eventually, that reformation exhausted itself or ran into sands. Why?

The poignant question was put to the present writer by a British journalist on a Churchill scholarship some years ago—in 1967 to be exact. We were both on a tour of the drought-stricken areas of Bihar and going by road from Patna to Gaya. On the way, in the hot afternoon, we stopped for an hour or more to go round the ruins of Nalanda, a great seat of Buddhist learning thirteen hundred years ago which had at times as many as 10,000 resident students, and which 'the Master of the Law', Hsuan-tsang, has described in one of his purple passages:

> The whole establishment is surrounded by a brick wall. One gate opens into the great college, from which are separated eight other halls, standing in the centre of the quadrangle. The richly-carved towers and fairy-like minarets cluster like hilltops; the upper storeys and observatories are lost in the morning mists. From the windows one sees the wind wreathing the clouds into various shapes, and from the soaring eaves one may observe the conjunction of the planets. Down below, the deep, transparent ponds bear on their surfaces the blue lotus, mingled with kanaka flowers, of a deep red colour; at intervals the amra groves throw a grateful shade over everything. All the outside courts, in which are the priests' chambers, are of four stages. The stages have carved and coloured eaves, pillars, and balustrades, and the tiled roof reflects the light in a thousand shades.

The glory which the Celestial saw vanished long ago. But even today the ruins of Nalanda have an unostentatious beauty which evokes something of the orderliness and symmetry of the Buddhist mind at its best. Visibly moved, my colleague asked: 'Why did it all have to come to an end?' Why, indeed? But this is not the place, nor the time, to find an answer to what remains a great and agonizing question-mark of Indian history. What is not in doubt, at least in my mind, is that when it ended, with it also ended a development—perhaps one should say, experiment—of immense significance and promise for humanity. And the pity of it . . .

ACKNOWLEDGEMENTS AND BIBLIOGRAPHY

It is pleasant to acknowledge one's debts—especially when acknowledgement happens to be the only means of repaying them. As this book was never intended to be a scholarly work, I did not think it necessary to burden it with footnotes. But omission to quote chapter and verse should not be taken to mean that I am any the less conscious of my debts. Among the ancients, I have drawn largely upon the anonymous compilers of the Pali and Sanskrit Canons; and in particular, I have found Asvaghosha's poem dealing with the early life of Gautama extremely rich in psychological detail. Among the moderns, I am greatly indebted to the works of the late Professor Rhys Davids to whose pioneer efforts we chiefly owe the English translation of invaluable Pali Texts; and I have constantly referred to Mr E. J. Thomas's *The Life of Buddha as Legend and History*. As regards my interpretation of Gautama's attitude to life, I should like to express my deep sense of gratitude to the late Professor Irving Babbitt. His *Rousseau and Romanticism* was of immense help to me in arriving at a clearer definition of some of the ideas I had vaguely formed in my own mind concerning the Buddha. Equally inspiring and fruitful was the reading of Sir Charles Eliot's *Hinduism and Buddha*—a work which impresses one not only by its vast scope, but also by its profound understanding. And although one may find it difficult to agree with everything that Mrs Rhys Davids says about Buddhism and the personality of its founder, a study of her works is an enriching experience in so far as it brings one into contact with a highly original and independent mind. No less enlightening are M. Louis de la Vallée Poussin's *Le Dogme et la Philosophie du Bouddhisme*, and

M. T. Stcherbatsky's monumental treatise on Buddhist Logic, as well as his various essays on the meaning of such difficult Buddhist concepts as Nirvana, Dharma, etc. M. René Grousset's *In the Footsteps of the Buddha* is perhaps the best guide to the Mahayana metaphysicians, and possesses the additional attraction of combining statement of facts with poetic realization of truth. Finally, I regard it as a privilege to acknowledge my indebtedness to the works of Professor S. Dasgupta and Dr Radhakrishnan. For, although there are many points on which I differ from their interpretations, their *Histories of Indian Philosophy* are among those basic books which are indispensable to any one who wishes to see the various phases of Indian thought in their historical perspective.

There are several reasons why I do not intend to draw up anything in the nature of an exhaustive bibliography, the principal reason being that excellent bibliographies are already available both in English and French. Moreover, I do not at all feel sure that it would serve any useful purpose to add a lengthy bibliography to a work which lays no claim to academic erudition. The general reader, I feel, seldom troubles himself with such appendages, and the serious student is usually better able to satisfy his curiosity elsewhere.

BIBLIOGRAPHY

GENERAL

Alabaster, Henry: *The Wheel of the Law: Buddhism Illustrated from Siamese sources*, 1871.

Babbitt, Irving: *Rousseau and Romanticism*, 1919.

Babbitt, Irving: *On Being Creative, and other Essays*, 1932.

Barua, B. M.: *A History of the Pre-Buddhistic Indian Philosophy*.

Barua, B. M.: *Prolegomena to a History of Buddhist Philosophy*, 1918.

Bell, Sir Charles: *The Religion of Tibet*, 1931.

Bhandarkar, D. R.: *The Inscriptions of Asoka*, 1920.

Blavatsky, Helena Petrovna: *The Secret Doctrine*, 2 vols., 1929.

Burnouf, Eugène: *Introduction à l'Histoire du Buddhisme Indien* (English translation by Winifred Stephens under the title: *Legends of Indian Buddhism*), 1876.

Cambridge Ancient History: Persian Empire and the West.

Carus, Dr Paul: *The Gospel of Buddha*, 1917.

Coomaraswamy, A. K.: *Buddha and the Gospel of Buddhism*, 1916.

Cranmer-Byng, Lionel: *The Vision of Asia*, 1932.

Csoma de Körös, Alexander: *Analysis of Kah-gyun, Asiatic Researches*, Vol. XX, 1820.

Dahlke, Dr Paul: *Buddhism and its Place in the Mental Life of Mankind*, 1927.

Dasgupta, S.: *History of Indian Philosophy*, 5 vols., 1922–55.

Dasgupta, S.: *Indian Idealism.*

David-Neel, Alexandra: *With Mystics and Magicians in Tibet*, 1931.

Davids, Mrs Caroline A. Foley Rhys: *Buddhism: A study of the Buddhist Norm*, 1912.

Davids, Mrs Caroline A. Foley Rhys: *Buddhist Psychology: An Enquiry into the Analysis and Theory of Mind in Pali Literature*, 1924.

Davids, Mrs Caroline A. Foley Rhys: *Buddhism, its Birth and Dispersal*, 1934.

Davids, Mrs Caroline A. Foley Rhys: *Sakya, or Buddhist Origins*, 1931.

Davids, Mrs Caroline A. Foley Rhys: *Will and Willer*, 1926.

Davids, Mrs Caroline A. Foley Rhys: Articles in *Hastings' Encylopaedia of Religions and Ethics* on the Buddhist view of Asceticism, Desire, Egoism, Happiness, Logic, Love, Music, *Paticca-Samupadda*, Perfection, Purification, Reality, Relations, Sacrifice, Self, Soul, etc.

Davids, T. W. Rhys: *Buddhism*, 1882.

Davids, T. W. Rhys: *Buddhist India*, 1911.

Davids, T. W. Rhys: *Lectures on the Origin and Growth of Religions as illustrated by the history of Buddhism*, 1881.

Davids, T. W. Rhys: Articles in *Hastings' Encylopaedia of Religions and Ethics* on Ananda, Buddhaghosa, Celibacy, Charity, Chastity, Devadatta, Dhammapala, Elder, Milandi, Moggallana, Tonsure, Wheel of the Law, Wisdom Tree, etc.

Dayal, Har: *The Bodhisattva doctrine in Buddhist Sanskrit Literature*, 1932.

Dowson, John: *A Classical Dictionary of Indian Mythology and Religion*, 1932.

Dutt, Sukumar: *Early Buddhist Monachism*, 1924.

Egoroff, Sophia: *Buddha, Sakyamuni—The Divine Socialist.*

Eliot, Sir Charles: *Hinduism and Buddhism*, 3 vols., 1921.

Erdmann, J. E.: *History of Philosophy*, 3 vols., 1893.

Garbe: *The Philosophy of Ancient India*, 1897.

Giles, H. A.: *The Travels of Fa-hsian, 399-414 A.D.*, 1923.

Goddard, Dwight: *The Buddha's Golden Path*, 1931.

Gour, Sir Hari Singh: *The Spirit of Buddhism*, 1929.

Grimm, G.: *The Doctrine of the Buddha, the Religion of Reason*, 1926.

Grousset, René: *In the Footsteps of the Buddha*, 1932.

Grousset, René: *The Civilizations of the East*, 4 vols., 1931.

Hardy, Robert Spence: *Eastern Monachism*, 1860.

Hayes, Helen M.: *The Buddhist Pilgrim's Progress, or Journey to the Western Paradise*, 1930.

Hearn, Lafcadio: *Gleanings in Buddha Fields*.

Herodotus: *History* (translation by Rawlinson, 2 vols.).

Hodgson, Bryan Houghton: Essays on the Language, Literature, and Religion of Nepal and Tibet, 1874.

Huc, Evarist Regis: *Travels in Tartary, Thibet, and China during the years 1844-46*, 2 vols., 1931.

I-Tsing: *The Buddhist Religion as Practised in India and Malay Archipelago, 617-695 A.D.*, translated by J. Takakusu, 1896.

Julien, Stanislas: *Voyages des Pèlerins Bouddhistes* (translated from the Chinese) 3 vols., 1853-8.

Keith, A. Berriedale: *Buddhist Philosophy in India and Ceylon*, 1923.

Knox, Robert: *An Historical Relation of Ceylon; together with Somewhat concerning several remarkable Passages of my life that hath happened since my deliverance from Captivity*, 1681.

Law, Bimla Charan: *Heaven and Hell in Buddhist Perspective*, 1925.

Levi, Sylvain: Articles in Hastings' *Encylopaedia of Religions and Ethics* on *Abhidhamma-Kosa*.

Lokanath, Bhikkhu: *Establishing the Sangha in the West*, 1932.

Masson-Oursel, Paul: *Ancient India and Indian Civilization* (written in collaboration with Helena de William-Grabowska, and Philippe Stern), 1934.

March, Arthur: *A Bibliography of Buddhism*, 1935.

Meyer, J. J.: *Sexual life in Ancient India*, 1935.

Massingham, H. J.: *The Heritage of Man*, 1929.

Mookerji, Radhakumud: *Asoka*, 1928.

Mookerji, Radhakumud: *Men and Thought in Ancient India*, 1881.

Müller, Max Frederick: *Selected Essays*, 2 vols., 1881.

Müller, Max Frederick: *Six Systems of Indian Philosophy*, 1899.
Okakura Kakuzo: *The Ideals of the East*, 1913.
Polo, Marco: *Travels*.
Przyluski, J.: *Le Bouddhisme*, 1932.
Radhakrishnan, Sir S.: *Indian Philosophy*, 2 vols., 1929-31.
Radhakrishnan, Sir S.: *An Idealist View of Life*, 1929.
Ragozin, Z. A.: *Assyria*, 1887.
Ragozin, Z. A.: *Chaldea*, 1886.
Rawlinson, George: *Phoenicia*, 1889.
Rawlinson, George: *Intercourse between India and the Western World from the Earliest Times to the Fall of Rome*, 1916.
Reade, Winwood: *Martyrdom of Man*.
Rost, Ernst R.: *The Nature of Consciousness*, 1930.
St Hilaire, J. Barthélemy: *The Buddha and his Religion*, 1914.
Saunders, Kenneth J.: *Buddhism*, 1929.
Senart, Emile: *Essai Sur La Légende du Buddha*, 1882.
Smith, Sir Grafton Elliot: *Human History*, 1934.
Smith, Sir Grafton Elliot: *The Diffusion of Culture*, 1933.
Smith, Vincent Arthur: *Asoka*, 1901.
Smith, Vincent Arthur: *Early History of India from 660 B.C. to Mohammedan Conquest*.
Stcherbatsky, Th.: *Buddhist Logic*, 2 vols., 1932.
Stcherbatsky, Th.: *The Conception of Buddhist Nirvana (including Candrakirti's commentary on Relativity)*, 1927.
Stcherbatsky, Th.: *A Treatise on Relativity by Nagarjuna, and Commentary thereon by Candrakirti*.
Stein, Sir Marc Aurel: *Innermost Asia*, 4 vols., 1928.
Suzuki, Daisetz Teitaro: *Studies in the Lankavatara Sutra*, 1930.
Thomas, E. J.: *The History of Buddhist Thought*, 1933.
Tucci, Giuseppe: *Some Aspects of the Doctrine of Maitreya and Asanga*, 1930.
Vallée-Poussin, Louis de la: *Le Dogme et la Philosophie du Bouddhisme*.

Vallée-Poussin, Louis de la: *The Way to Nirvana*, 1917.
Waley, Arthur: *The Way and Its Power*, 1935.
Waley, Arthur: *Did Buddha die of eating Pork?* An article in *Mélange Chinois* (Bruxelles), pp. 343-54.
Winternitz, Moriz: *A History of Indian Literature*, 2 vols., 1927.

BIOGRAPHICAL

Arnold, Sir Edwin: *The Light of Asia*, 1878.
Bigandet, Bishop of Ramatha: *The Life and Legend of Gaudama, the Buddha of the Burmese*, 1880.
Brewster, E. H.: *The Life of Gotama Buddha*, 1926.
Davids, Mrs Caroline A. Foley Rhys: *Gotama the Man*, 1928.
Oldenberg, Hermann: *Buddha, his life, his doctrine, his order*, 1929.
Saunders, Kenneth J.: *Gotama Buddha*, 1922.
Thomas, E. J.: *The Life of Buddha: as Legend and History*, 1927.

SACRED BUDDHIST TEXTS

(Modern translations of canonical and post-canonical works)

Babbitt, Irving: *Dhammapada* (translation with an introduction), 1935.
Barnett, Lionel D.: *The Path of Light; A Manual of Mahayana Buddhism translated from the Bodhicharyavatara of Santi-Deva*, 1909.
Beal, Rev. Samuel: *A Catena of Buddhist Scriptures from the Chinese*, 1871.
Chalmers, Lord: *Further Dialogues of the Buddha translated from the Pali Majjhima Nikaya* in 2 vols., 1926-7.
Davids, Mrs Caroline A. Foley Rhys: *Buddhist Birth Stories*, 1925.
Davids, Mrs Caroline A. Foley Rhys: *Psalms of the early Buddhists*, 2 vols., 1909-13.

Davids, Mrs Caroline A. Foley Rhys: *The Book of the Kindred Sayings*, see Woodward.

Davids, T. W. Rhys: *Buddhist Suttas* (Vol. XI of the *Sacred Books of the East*, containing *Mahaparinibbana Sutta, Dhammacakkappavatana Sutta, Tevigga Sutta, Mahasudassana Sutta, etc.*), 1881.

Davids, T. W. Rhys: *Dialogues of the Buddha, being a translation of the Digha Nikaya*, in 3 vols. (Vols. II and III translated in collaboration with Mrs Rhys Davids).

Davids, T. W. Rhys: *The Questions of King Milinda* (Vols. XXXV and XXXVI of the *Sacred Books of the East*), 1890.

Davids, T. W. Rhys: *Vinaya Texts*, 3 vols. (Vols. XIII, XVII, and XX of the *Sacred Books of the East*); also see Oldenberg.

Davids, T. W. Rhys: *Pali English Dictionary*, 4 vols.

Fausböll, Victor: *Sutta Nipata* (*Sacred Books of the East*, Vol. X), 1898.

Geiger, Wilhelm: *Culavamsa*, 2 vols., see Rickmers.

Kern, Johan Hendrik: *The Saddharma-Pundarika, or the Lotus of the True Law* (*Sacred Books of The East*, Vol. XXI), 1884.

Müller, Max Frederick: *Buddhist Mahayana Sutras* (Vol. XLIX of the *Sacred Books of the East*) containing the *Buddha-Charita of Asvaghosha, the larger Sukhavati-Vyuha*, etc.

Obermiller, E.: *Bu-Ston, A History of Buddhism* (*the Jewelry of Scripture*) with introduction by Th. Stcherbatsky.

Oldenberg, Hermann: *Vinaya Texts*, see T. W. Rhys Davids.

Rickmers, C. Mabel: *Culavamsa*, being the more recent part of *Mahavamsa*, 2 vols. (with Wilhelm Geiger), 1929.

Thomas, E. J.: *Buddhist Scriptures: A Selection*, 1913.

Woodward, F. L.: *The Book of the Gradual Sayings* (*Anguttara Nikaya*), 2 vols.

Woodward, F. L.: *The Book of the Kindred Sayings* (*Samyutta Nikaya*), in 2 vols. (with Mrs Rhys Davids).

BUDDHIST ART

Aneski, Masahara: *Buddhist Art in its relation to Buddhist Ideals*, 1916.

Codrington, K. de B.: *Ancient India, from the Earliest Times to the Guptas*—Prefatory essay by W. Rothenstein, 1926.

Coomaraswamy, A. K.: *Early Indian Iconography in Eastern Art*, Vol. I.

Ferguson, J.: *Tree and Serpent Worship*, 1873.

Foucher, A.: *The Beginning of Buddhist art and other Essays*, 1919.

Havell, E. B.: *Ideals of Indian Art*, 1920.

Herringham, Lady: *Ajanta Frescoes*, 1915.

Kramrisch, Stella: *Buddhist Art in India*, 1928. (*Buddhist India*, pp. 26-33.)

Krom, N. J.: *The Life of Buddha on the Stupas of Borobudur*, 1926.

Marshall, Sir John: *A Guide to Taxila*.

Stein, Sir Marc Aurel: *The Thousand Buddhas: Ancient Buddhist Paintings from the Cave Temples of Tung-Huang on the Western Frontier of China, introductory essay by Laurence Binyon*, 1921.

Vogel, J. Ph.: *Buddhist Art in India*.

Vogel, J. Ph.: *Catalogue of the Museum of Archaeology at Sarnath*, 1914.

Waley, Arthur: *Zen Buddhism and its Relation to Art*, 1922.

Waley, Arthur: *An Introduction to the Study of Chinese Painting*, 1923.

Waley, Arthur: *A Catalogue of the Paintings recovered from Tung-Huang*, 1931.

Yazdani, J.: *Ajanta*, colour and monochrome reproduction with introduction by Laurence Binyon.

BUDDHISM AND CHRISTIANITY

Carpenter, J. Estlin: *Buddhism and Christianity: A Contrast and a Parallel*, 1923.

Dasa, Philangi: *Swedenborg the Buddhist*.

Edmunds, A. J.: *Buddhist and Christian Gospels compared*, 2 vols., 1908-9.

Hayes, Will: *How the Buddha became a Christian Saint*, 1931.

Jacobs, Joseph: *Barlaam and Josaphat, or English lives of Buddha*, 1898.

Lillie, Arthur: *Buddhism in Christendom, or Jesus the Essene*, 1887.

Monier-Williams: *Buddhism in its connection with Brahmanism and Hinduism, and its contrast with Christianity*, 1890.

Renan, Ernest: *Life of Jesus*.

Robertson, J. M.: *A Short History of Christianity*.

Robertson, J. M.: *Pagan Christs*.

Saunders, Kenneth J.: *Christianity and Buddhism* (Proceedings of the International Missionary Council, 1928).

Streeter, B. H.: *The Buddha and the Christ*, 1932.

INDEX

Abhaya 224
Aegean League, the 28
Africa 15, 233
Aholah and Aholibah 37
Ahuramazda 29
Aiterya Upanishad 65
Ajanta 291-5, 298, 318
Ajatasattu 67, 81, 231-2, 237, 239-41, 243, 249-50
Ajita 247
Alabaster, Henry 129
Alara Kalama 154-5, 168, 180
Alavi 215
Al Biruni 313
Alexander the Great 27, 29, 35, 49, 350
Allakappa 250
All's Well That Ends Well 201
Ambapalika 108, 109, 243
Amor, St 106
Amravati 296, 298, 300
Amytis 26
Ananda 73-4, 128-9, 206, 210-11, 215-16, 221-3, 232, 242, 244-8, 336
Anattalakkhana-sutta 277
Aneski, M. 322
Angulimala 215
Anguttara Nikaya 86
Anoma 206
Antioch 313
Antiochus 313
Antisthenes 350
Anupiya 206
Anuruddha 206
Apollo 59
Arabian Sea, the 235
Aragon, Louis 305, 355
Archelaus, Acts of 47, 314
Archelaus, Bishop 47
Aretino 299
Aristagoras 34
Aristophanes 350
Aristotle 281, 343
Arnold, Edwin 322
Asanga 288, 301
Asia 15, 27, 34, 233
Asia Minor 16, 27, 328
Asita 53, 82
Asoka 46, 313
Aspasia 350
Assaji 195-6, 227
Asshur 23, 38
Assyria 24, 25
Asvaghosha 87-9, 91-2, 111, 114-16, 122, 124-5, 139-40, 162, 288, 359
Athenaeus 350
Athens 28, 35, 107, 313, 350
Augustus 313
Axel 95

Babbitt, Irving 316, 355, 359
Bab-el-Mandeb 235
Babylon 23, 25-6, 33, 35, 54
Bach 190
Bactria 31, 278
Barlaam and Josaphat 318
Barton, Bruce 324-5
Baudelaire 95, 130, 256, 258
Baumgarten 299
Beauvais, Vincent de 319
Behistun inscription 235
Being Creative 316
Belshazzar 26
Bergson 283, 352
Berkeley, Bishop 285
Beluva 244
Benares 73-4, 87, 107, 121, 180-3, 187-8, 247, 277
Benda, Julien 317
Beyond The Mexique Bay 299
Bharadvaja 214
Bhaddakaccana 92-3, 136
Bhagavat Gita 342
Bharut 298
Bhima 105
Bidatsu 312

Bigandet, Bishop 320-1
Bimba 92
Bimbisara 81, 109, 154, 193, 208, 211, 222, 224, 237, 239, 241
Binyon, Laurence 291, 296
Birmingham, Bishop of 320
Bodhidharma 268, 301
Bodhisattva Padmapani 295, 302
Boethius 142
Book of The Great Decease 242
Book of The Nines 127
Borobudur 296, 302, 318
Brahma 63, 65, 102, 146, 148, 152, 199, 268, 275-6
Brahmajala-sutta 150, 266, 279
Brahmana of the Hundred Paths, the 146
Bruce, Rev. 342
Bubastis 32
Buddhaghosa 217-9
Buddha-Karita 87
Buddhism 61, 84, 198, 279, 280, 291, 293, 307-10, 313, 315-16, 320-1, 325-6, 331, 334, 342, 347
Bulis 250
Burnouf, Eugène 314-15

Callimachus, Polemarch 28
Cambrai, Guy de 319
Cambridge Ancient History 31
Canaan, Sons of 16
Canton 301
Carpenter, Frederick Ives 342
Caspian Sea, the 58, 234
Chandra Gupta 29, 54
Channa 138, 139, 140-1, 154
Charvakas 148
China 24-5, 27, 29, 33, 37, 268, 302, 322, 325
Chincha 212
Clement of Alexandria 313
Cleopatra 15
Columbus 35
Comte 267
Confucius 27, 37, 325
Constantine 313
Coomaraswamy, A. K. 292, 310
Counsels and Maxims 123
Croesus, King of Lydia 255
Csoma de Körös, Alexander 314
Ctesias 31
Cullavagga 298
Cyrene 313
Cyrus, The Great 26, 29, 31, 233, 235-6

Dalai Lama 320
Daniel 26
Dante 307
Darius 27, 31-2, 35, 233, 235
Dasgupta, S. 360
David, Son of 17
Davids, Mrs Rhys 170, 286, 359
Davids, T. W. Rhys 46, 69, 83, 162-3, 221, 278-9, 282, 315, 317-18, 320, 359
De Imitatione Christi 297
Delphi 285, 349
De Musset 345
Democritus 278, 286
Descartes 270, 283, 285
Devadatta 78-9, 80-1, 92-3, 206, 239, 240-1
Dewey, John 276
Dhammapada 226, 258, 283
Diaz 35
Diganaga 288
Digha-Nikaya 282
Diogenes 350
Dirghatamas 106
Dona 250
Dostoievsky 256
Dubois, Abbé 108
Duhn, Frederick 105

Ecclesiastes 119, 190
Edmunds, A. J. 322
Egoroff, Sophia 334
Egypt 15, 26, 32-3, 35, 40, 144, 234-5, 313
Eleatics 349
Elephanta 300

Eleusinian Mysteries 349
El Greco 329
Eliot, Sir Charles 169, 224, 330-1, 359
Eliot, T. S. 45, 258
Ellora 292
Emerson, Ralph Waldo 342
Empedocles 40, 255
Engels 265
England 55
Ephesus 283, 291
Epirus 313
Erdmann 19
Erythrean Sea, the 35
Ethiopia 234, 318
Europe 16, 27, 34, 62, 67, 90, 233, 337, 348
Eyeless in Gaza 283
Eysinga, Van der Bergh Van 322

Fact and Faith 316
Fa Hsian 72-3, 240-1, 301
Fa-men 346
Fausböll, Victor 315
Feuerbach 175
Fichte 285
Finot 297
Fire Sermon, the 190, 258-9
Foundation of The Kingdom of Righteousness 262
France 95
France, Anatole 333
Francis, St, of Assisi 326
Freud 258

Galilee 324-5, 330
Gandhara 59, 222, 235
Garbe 323
Garland of Flowers 287
Gauguin, Paul 300
Gautama, the Buddha (also Blessed One, Bodhisattva, Buddha Amida, Enlightened One, Exalted One, Light of the World, Sakya Muni, Siddhartha), 26-7, 38, 40-1; problem of historicity, 45-6; claims of Virgin Birth, 47-8; Marks of Greatness and other circumstances of birth, 47-53; as the Great King of Glory, 54, 57, 59, 63, 67, 69, 70-4, 77; rivalry with Devadatta, 78-81; name-giving ceremony, 82-3; early life, 86-90; marriage, 90-4; Crisis and Renunciation, 118 *et seq.*; austerities, 156-8; combat with Mara, 161; some interpretations of Enlightenment, 162 *et seq.*; First Sermons, 185-6; Foundation of the Brotherhood, 187-9; the Fire Sermon, 190-1; visit to Rajagaha, 192-8; opposition of the Brahmanical and Patriarchal Order, 198-203; return home, 203-7; visit to Savatthi, 208-9; death of Suddhodana, 210; Foundation of the Order of Female Mendicants, 211-12; first dissension, 213-14; daily routine, 216-21; his circle, 221-5; his sense of humour, 230-2; last years, 237-8; Devadatta's Schism, 239-41; last days, 242-8; trouble over his ashes, 249-50; the Truth of Pain, 253-6; the Truth of the Cause of Pain, 257-9; the Truth of the Cessation of Pain, 260-1; The Way as a Paradox, 262 *et seq.*; his rejection of the metaphysical approach, 272-4; his agnosticism, 274-7; his denial of soul, 277-81; his empiricism, 283-8, 299, 303, 307-14; his Canonization as St Josaphat, 318 *et seq.*; comparison with Jesus, 322-33; as a social and religious reformer, 334-9; his humanism, 339-42; his critics, 342-53; summing up, 355-9
Gaya 182, 189, 191-2
Genesis 19
Germany 106
Giorgione 300

Givaka Komarabhakha 89, 223-4, 241
Goethe 339, 352
Gopa 92
Gore, Bishop 320
Gorgias 281
Gradual Sayings 127
Gray, G. Buchanan 26
Greece 21, 25, 56, 106, 151, 349-50
Grousset, René 167, 284, 287, 298, 301, 360

Hadda 318
Haldane, J. B. S. 315
Hamlet 118, 256, 332
Han-Yü 347
Harsha 301
Hardy, Spence 315
Hari Singh Gaur 313
Havell, E. B. 301
Hayes, Will 311-12
Hecuba 16
Hellas 33
Hellenes 18
Hellenism 19, 349
Hera 349
Heraclitus 22, 278, 283
Hermes 59
Herodotus 30-1, 235
Hesiod 21
Hibbert Journal, the 333
Hinayana Buddhism 273, 313
Hinduism and Buddhism 359
Hippias 281
Hiram 16
Histiaeus 34
Historical Relation of Ceylon 314
Hodgson, Brian Houghton 314
Holy Inquisition, the 321
Hopkins, Gerald Manly 286
Hsia Kuei 303
Hsien Tsung 346
Hsuan-tsang 72, 288, 301
Hume, David 273, 285
Huxley, Aldous 283, 298-9
Huxley, Thomas 308

I-Ching 301
India 24-5, 29, 31-3, 38, 41, 49, 52, 54-9, 61, 67, 69, 72-3, 89, 91, 95-6, 99, 102, 105-6, 108, 114-15, 143, 151, 181-2, 235-6, 241, 278, 281, 291, 296, 299, 302, 310, 313, 318, 323, 326, 327, 331, 337, 349, 355
Inge, Dean 324, 352
Introduction à l'histoire du Buddhisme Indien 314
Ionia 34
Isaiah 11, 293, 330

Jacobs, Joseph 318
James, William 317
Japan 302, 322
Jataka 46, 126
Jaxartes 234
Jehovah 280
Jeremiah 24, 330
Jerome, St 47, 314
Jerusalem 24, 35, 120, 321, 328
Jesting Pilate 298
Jesus (Christ) 46, 47, 78, 314, 322-33, 338, 355
Job 40, 255, 330
John, the Monk 318
John, St 251
Journey To The Light 297, 301
Judas Iscariot 78
Justinian Code 321

Kabir 255
Kakkayana 247
Kama Sutra 64, 104
Kanishka 87, 114
Kant 285, 286, 352
Kapila 38, 139, 146-7, 155, 277
Kapilavastu 46-7, 52, 54, 56, 57, 59, 71-4, 117, 136, 138, 139, 140, 146, 155, 203-5, 210-11, 215, 238, 250, 310, 319-20
Kassapas 189

Kautsky, Karl 324, 334
Keith, Berriedale 347-8, 351-2
Kern, H. 167, 323
Kipling, Rudyard 317
Kisa Gotami 137-8
Knox, Robert 314-5
Kondanna 82
Kosala 56, 73, 208, 222, 237-8
Kosambi 74, 211-13, 247
Krishnamurti 132
Krom 302
Kuan-Yin 84
Ku K'ai-chi 302
Kunda 245-6
Kusavati 86, 117
Kusinara 73, 243, 246-7, 249-50, 317
Kyropaedia 235

La Loubere, Simon de 314
Lalita-vistara 47, 53, 89, 319
Language and Wisdom of the Hindus 345
Lao Tzu 37
Lawrence, D. H. 256, 333
Leibnitz 285
Lenin 335
Leonardo da Vinci 16
Leucippus 278
Lhasa 320
Licchavis 56, 210, 217, 236, 250
Licht 101
Life of Buddha as Legend and History 359
Light of Asia 322
Li Shang-yin 301
Locke 285
Loisy 323
Lokayatikas 148, 179, 198, 254
Louis XIV 314
Lung-mên 295, 318
Luther 335

Macdonell 103
Madhyamika Doctrine 166, 169, 287
Magandiya 212-13
Magas 313
Mahabharata 103, 105
Mahanama 238
Mahanidana-sutta 282
Mahapajapati 205, 210
Magadha 56, ,73, 89, 154, 156, 191, 197, 201, 203, 208, 217, 222-4, 237-8, 249-50
Mahasaccaka-sutta 164
Mahasudassana Sutta 54, 86
Mahavira 153
Mahayana Buddhism 78, 166, 273, 284, 286, 300-1, 310
Maitri Upanishad 65
Majjhima-Nikaya 177, 181, 185
Makkhali Gosala 150, 153, 183, 247
Ma Lin 303
Mallas 73, 243, 246, 249-50
Malunkya 171-2
Mani, the Heretic of Ecbatana 47, 313
Manu 57, 199, 221
Manu, the code of 68
Mara 161, 162, 167
Marathon 28
Marlowe 336, 340
Marx, Karl 175
Massingham, H. J. 309
Maya 47-8, 49, 50, 52
Megasthenes 69, 95, 181
Meleager 59
Memphis 33
Mesopotamia 27, 144
Meyer, J. J. 101, 103, 106
Michael Angelo 17, 296, 298
Migara 225
Miletus 34, 41
Milinda (Greek Menander) 278
Miltiades 28
Milton 162
Moggallana 194, 196, 216, 223, 239, 241
Mohammed 27
Monier-Williams, Professor 59
More, Paul Elmer 343

Müller, Max 20, 317, 319-20
Murry, J. Middleton 78

Nagarjuna 166, 287, 300
Nagasamala 215-16
Nagasena 278
Nahum, the Elkoshite 23
Nakhs-i-Rustam Inscription 235
Nalanda 230
Nanda 205-6
Nara 296, 302
Nebuchadnezzar 23, 25, 26
Necho 25, 32, 35
Nepal 56, 59, 314
Newton 174
Nietzsche 259-60, 261, 317, 333
Nineveh 23, 33
Nouvelle Héloïse 345

Okakura Kakuzo 307
Oldenberg, Herman 208, 233, 236, 315, 327
Olympias 49
Osiris 21, 54, 95
Othello 256, 356
Otto, Dr 331

Palestine 323
Panthakas 221, 229
Papini 324
Paraçara 91
Paradise Regained 162
Pasenadi 72, 89, 208, 215, 222, 237, 241
Patliputtra (Pataligama) 69, 238, 243
Pava 245, 250
Pericles 350
Pelliot, M. 302
Peloponnesian Wars 28
Persepolis Inscription 235
Persia 26, 27, 30-2, 54, 235
Phaedo 233
Philip of Macedon 49
Pischel 322
Plancy, Collin de 318
Plato 107, 233, 259, 278, 349-50
Pliny 77
Pokkharadi 223
Pollaiuolo 296, 298
Polo, Marco 314
Pontius Pilate 325
Preacher, the 17, 40, 120
Preuss, Dr 349
Przyluski, Jean 334-5
Ptolemy 313
Pythagoras 37
Pythagoreans 349

Radhakrishnan, Sir S. 280, 333, 360
Ragozin 23
Rahula 136, 138, 205-6
Rajagaha 73, 107, 108-10, 154, 190-4, 201, 206-10, 215, 217, 239, 243
Reade, Winwood 233
Red Sea, the 32, 35, 235
Reichelt, Dr 325, 330
Remy de Gourmont 90
Renan 316, 331
Renoir 300
Rhodes, Cecil 317
Richards, I. A. 220
Rig-Veda 19, 145
Rochester 345
Rome 95
Rousseau 116, 339-40
Rousseau and Romanticism 359
Rubens 300
Russell, Bertrand 352
Russia 29

Sakyas 57, 59, 72-3, 79, 210, 217, 238, 250
Salavati 109
Sama Veda 272
Samyutta-Nikaya 160, 175
Sanchi 296, 298, 300, 301
Sanjaya 194, 196-9, 203
Sankra 259, 278
Santayana, George 282

Santideva 297-8
Saracos 23
Sardes 31
Sariputta 194-5, 205, 216, 223, 227, 230-1, 241
Sarnath 59
Saunders, K. J. 321
Savatthi 73, 107, 208, 214-17, 225, 296
Schiller 230
Schlegel, Frederick 345
Schopenhauer 123, 133, 260, 271, 285, 345, 352
Schweitzer 355
Seleucus Nicator 69
Senart, Emil 45, 315, 323
Sermon on the Mount 185, 323
Seydel 322
Shaburkan 313
Shakespeare 46, 258
Sidon 16, 33
Signorelli 296, 298
Silabhadra 288
Simeon 53, 322
Sistine Chapel 293
Siva 57, 181, 349
Skylax 35, 235
Smith, Elliot 19
Socrates 46, 107, 256, 285, 348-9, 350-1
Solomon 17, 124
Solon 25, 255
Sonadanda 223
Song of Songs 124, 190
Sophocles 350
Sotthiya 160
Sparta 28, 350
Speculum Historiale 319
Spinoza 281
Stcherbatsky, T. 315, 360
Stchoukine, Ivan 292
Stein, Sir Aurel 302
Story of the Lineage 47, 50
Strabo 70, 313
Strachey, Lytton 45, 352
Streeter, Canon 324, 330
Suddhodana 47, 49, 50, 53, 57, 79, 83-4, 86, 88, 89, 91-3, 140, 204-6, 210
Sudatta (Anathapindika) 208-10, 223
Suez 32, 35, 320
Suppabuddha 92-4, 214-15
Susa 31, 33-5
Suzuki 315
Synoptic Gospel 324

Taittiriya Upanishad 65
Takhasila (Taxila) 88-9, 222-3
Tang Hsi 37
T'ao Han 302
Tao Te King 37
Tempest 271
Terebinthus 47
Tevigga Sutta 275
Thales 37, 41
Thalheimer, August 336
Thera Gatha 221
Thomas à Kempis 297
Thomas, E. J. 57, 77, 315, 359
Thoreau 322
Titian 300
Torquemada 321
Townsend, Meredith 317
Tun-Huang 295, 302
Turnour, George 315
Tyre 16

Udayin 124
Uddaka Ramaputta 155, 169, 180
Udena 213
Uggasena 210
Upaka 182-3
Upali 206, 221, 223
Upanishads 65, 329
Uppala-Vanna 80
Ur of the Chaldees 16, 33
Uttara Mimansa 282

Vajjians 217, 236-8, 243
Vallée Poussin, Louis de la 171, 268, 315, 359

Vamsas 217
Vasco da Gama 35
Vasettha 63, 275-6
Vasiliev 315
Vasistha 91, 124, 201
Vasubandhu 286, 288, 300
Vatsyayana 104
Vedas 69, 149, 275-6, 337
Vesali 108, 109, 210, 243-5, 250
Villiers de l'Isle Adam 95
Vinaya, the 70, 108, 185, 224, 226, 230
Visakha 224, 296
Vishnu 66, 349
Voltaire 276
Von Hartmann 352
Voragine, Jacques de 319
Vyasa 282

Wachtmeister, Axel Raoul 318
Wells, H. G. 175
Werther 345
Whitehead, A. N. 331
Woodward, E. L. 315
Works and Days 21
Wu Tao-tzu 302

Xenophon 235

Yajnavalkaya 91, 201, 240
Yamato 312
Yasa 187-8
Yasodhara 92
Yih-King 21-2
Yogacara Doctrine 286, 288

Zamanochegas 313
Zedekiah 24
Zen Buddhism 268
Zeno 22, 349
Zetland, Lord 308
Zeus 349
Zeus-Ammon 49
Zoroaster 77, 236, 314